Lecture Notes in Civil Engineering

Volume 301

Lecture Notes in Civil Engineering (LNCE) publishes the latest developments in Civil Engineering—quickly, informally and in top quality. Though original research reported in proceedings and post-proceedings represents the core of LNCE, edited volumes of exceptionally high quality and interest may also be considered for publication. Volumes published in LNCE embrace all aspects and subfields of, as well as new challenges in, Civil Engineering. Topics in the series include:

- Construction and Structural Mechanics
- Building Materials
- Concrete, Steel and Timber Structures
- Geotechnical Engineering
- Earthquake Engineering
- Coastal Engineering
- Ocean and Offshore Engineering; Ships and Floating Structures
- Hydraulics, Hydrology and Water Resources Engineering
- Environmental Engineering and Sustainability
- Structural Health and Monitoring
- Surveying and Geographical Information Systems
- Indoor Environments
- Transportation and Traffic
- Risk Analysis
- Safety and Security

To submit a proposal or request further information, please contact the appropriate Springer Editor:

- Pierpaolo Riva at pierpaolo.riva@springer.com (Europe and Americas);
- Swati Meherishi at swati.meherishi@springer.com (Asia—except China, Australia, and New Zealand);
- Wayne Hu at wayne.hu@springer.com (China).

All books in the series now indexed by Scopus and EI Compendex database!

N. A. Siddiqui ·
Akmalov Shamshodbek Baxtiyarovich ·
Abhishek Nandan · Prasenjit Mondal
Editors

Advances in Waste Management

Proceedings of the International Conference on Advances and Innovations in Recycling Engineering (AIR-2021)

Springer

Editors
N. A. Siddiqui
Centre of Excellence in Occupational
Health, Safety, Fire and Environment
GD Goenka University
Sohna, Haryana, India

Akmalov Shamshodbek Baxtiyarovich
Department of Civil Engineering
Tashkent Institute of Irrigation
and Agricultural
Tashkent, Uzbekistan

Abhishek Nandan
Department of Civil Engineering
University of Petroleum and Energy Studies
Dehradun, Uttarakhand, India

Prasenjit Mondal
Centre of Excellence in Occupational
Health, Safety, Fire and Environment
GD Goenka University
Sohna, Haryana, India

ISSN 2366-2557 ISSN 2366-2565 (electronic)
Lecture Notes in Civil Engineering
ISBN 978-981-19-7505-9 ISBN 978-981-19-7506-6 (eBook)
https://doi.org/10.1007/978-981-19-7506-6

This Springer imprint is published by the registered company Springer Nature Singapore Pte Ltd.
The registered company address is: 152 Beach Road, #21-01/04 Gateway East, Singapore 189721, Singapore

Preface

Waste management rules in India are based on the principles of "sustainable development", "precaution" and "polluter pays". These principles mandate municipalities and commercial establishments to act in an environmentally accountable and responsible manner—restoring balance if their actions disrupt it. The increase in waste generation as a by-product of economic development has led to various subordinate legislations for regulating the manner of disposal and dealing with generated waste are made under the umbrella law of the Environment Protection Act, 1986 (EPA). Specific forms of waste are the subject matter of separate rules and require separate compliances, mostly in the nature of authorizations, maintenance of records and adequate disposal mechanisms.

Waste Management (industrial and domestic) is one of the major concerns of modern society due to the ever-expanding volume and complexity of discarded domestic and industrial waste. It has become one of the most significant environmental issues, particularly in developing countries. In India, the approach toward solid waste management is largely unscientific. Where even in India, today a large portion of waste is being dumped indiscriminately on the outskirts of towns or cities, without any prior treatment, which is leading to groundwater contamination and an increase in pollution across the country.

If the Solid Waste Management Rule is properly implemented and is also adopted by the people, it can simply transform the waste management system in India. The amount of high-calorific waste is increasing in India and the country is gradually trying to adopt waste segregation at source, these two factors are crucial in running waste-to-energy projects, apart from that, there exist waste-to-energy technologies that can process unsegregated and high moisture-low calorie waste too and that can be very useful for a country like India as in many cases bringing a quick change in people's behavior is not easy and something needs to be done urgently.

Through this publication, the reader can update himself/herself with the advances in water pollution monitoring and control, and inform on related opportunities and challenges. This volume presents select papers on Advances in waste management which were presented at the International Conference on Advances and Innovations in Recycling Engineering (AIR-2021), October 21–22, 2021, organized by the

University of Petroleum and Energy Studies, Dehradun (UPES) India, in collaboration with Department of Chemical Engineering, Hindustan Institute of Technology and Science (HITS), Chennai, India and The Department of Hydrology and Hydrogeology, Tashkent Institute of Irrigation and Agricultural Mechanization Engineers, Uzbekistan. The conference was attended by leading academic scientists, leading engineers, policy makers, budding scholars and graduate students. The contribution from the authors cover topics ranging from waste management to clean development mechanism that would help tackle the problem of Recycling Engineering and Waste Management and ensure access to eco-friendly techniques for waste management to be adopted to ensure sustainable development.

Dehradun, India Dr. Abhishek Nandan

Contents

About the Editors

Dr. N. A. Siddiqui is currently working as Professor and Director at Centre of Excellence in Occupational Health, Safety, Fire and Environment, GD Goenka University, Gurgaon, India. He pursued his postgraduate degree in environmental science and doctorate in environmental biology. In addition, he also has an International Diploma in Occupational Health (IDip. NEBOSH, UK), Diploma in Industrial Safety, and Postgraduate Diploma in Environmental Impact Assessment. His research interests lie in environmental impact assessment. He specializes in the areas of safety management, occupational health and safety, environmental pollution, environmental monitoring, and control techniques. He was also associated with Health, Safety and Environment Department of ICEM College, Muscat, Oman, which is affiliated to the University of Central Lancashire, UK. He has more than 170 research papers to his credit and has participated in several national and international conferences. He has authored 14 books and has guided more than 200 M. Tech. thesis and 21 Ph.D. theses.

Dr. Akmalov Shamshodbek Baxtiyarovich is currently working as an Associate Professor and International Grants Coordinator, in the Hydrology and Hydrogeology Department at the Tashkent Institute of Irrigation and Agricultural Mechanisation Engineers National Research University, Uzbekistan. Dr. Baxtiyarovich pursued his postgraduate and master's degree in environmental science and doctorate in hydraulics and engineering hydrology. Currently, he is providing Ds researches in technical science. In addition, he also working as an International Grants Coordinator in educational and research grants with many foreign universities, and as a national project coordinator in many research projects. His research interests lie in water resources, hydrology and GI. He specializes in the areas of water resources management, remote sensing and GIS in hydrology, environmental protection, climate change, water monitoring, and control techniques. He has over 80 research papers to his credit and has participated in several national and international conferences. He has authored six books and has guided five graduate, one master theses and one Ph.D. theses.

Dr. Abhishek Nandan is assistant professor at University of Petroleum and Energy Studies (UPES), Dehradun. His areas of interest are ambient and indoor air quality improvement techniques, drinking water and industrial effluent treatment strategies, conversion of waste to wealth and sustainable development techniques. He has worked in several internal and external funded projects as investigator. He has published more than 50 international journal papers and edited book chapters. His major patents includes Cancer Risk Assessment tool using Python Language, Indoor air pollution control by application of IoT and Water Filtration System using bio-adsorbents.

Dr. Prasenjit Mondal is currently working as Associate Professor at Centre of Excellence in Occupational Health, Safety, Fire and Environment, GD Goenka University, Gurgaon, India. His areas of interest are drinking water and industrial effluent treatment strategies, conversion of waste to wealth, ambient and indoor air quality improvement techniques, industrial hazard and risk analysis. During his Ph.D. work, he developed a low budget and highly efficient drinking water filtration system using organic adsorbent. He has worked in several internal and external funded projects as investigator. He has published more than 30 international journal papers and edited book chapters. His major patents includes Cancer Risk Assessment tool using Python Language, Indoor air pollution control by application of IoT and Water Filtration System using bio-adsorbents.

Assessment of Groundwater Using Water Quality Index (WQI) at Saharanpur City, Uttar Pradesh (West), India

Abhishek Tyagi, Satayajeet Arya, Abhishek Nandan, and Prasenjit Mondal

Abstract Groundwater is a natural resource and plays a vital role in our life. The study was carried out to access the groundwater quality of Saharanpur district, Uttar Pradesh. The assessment study was formulated in WQI to understand more about groundwater quality in a single term. Water Quality Index (WQI) summarizes numerical equations in a single term to understand better the quality of water. It is also helpful in determining the valuable rating of water quality and appropriate technique for its treatment. It also communicates information about water quality to the public and legislative decision-makers. In the present study, groundwater samples were collected from different locations, and WQI has been computed using seven parameters viz., pH, Total Hardness, Total Alkalinity, Electrical Conductivity, Calcium, Magnesium and Chloride. The result shows that WQI for all the locations were higher than 100 and in some locations, it was more than 200, which means the water quality is extremely poor and not drinkable.

Keywords Water quality index · Groundwater quality · Water analysis

1 Introduction

Groundwater is a natural dynamic renewable resource with consideration of all others [1]. Its availability in adequate quantity is very important for human life and other purposes. Human life depends, in direct (for drinking) and indirect ways (like cooking, washing, bathing, etc.), on fresh water. Groundwater is the most crucial source of potable water throughout the world [2]. It is generally consumed

A. Tyagi (✉)
Environmental Management Division, Central Pulp and Paper Research Institute, Saharanpur, India
e-mail: abhishektyagi34@gmail.com

S. Arya
Sri Sri University, Cuttack, Odisha, India

A. Nandan · P. Mondal
University of Petroleum and Energy Studies (UPES), Bhidoli, Dehradun, Uttrakhand, India

N. A. Siddiqui et al. (eds.), *Advances in Waste Management*, Lecture Notes in Civil Engineering 301, https://doi.org/10.1007/978-981-19-7506-6_1

by drinking, washing, preparing food and so forth. Groundwater defilement due to anthropogenic exercises is a worldwide issue for domain researchers and policymakers. Among the anthropogenic exercises, industrialization, urbanization, solid waste unloading, present-day rural and so forth assume a huge part in tainting of freshwater aquifers [3–8]. But presently due to a lack of discipline and weak legislations toward conservation, the quality and quantity of water became polluted and spoiled. Consequently, the number of water-borne diseases which cause health hazards has increased [9]. Nonetheless, the greater part of the investigations connected with groundwater quality examinations have been completed in the eastern or focal district of Uttar Pradesh, and there is an earnest need to lead such a review in the western locale. Saharanpur district falls under the Hindon River catchment [2]. That's why it is necessary to monitor the quality of groundwater regularly to observe the demand and level of pollution in it.

The present study mainly focused on the physiochemical analysis of groundwater samples of different locations and formulated the results in the WQI to conclude the exact quality of groundwater whether drinkable or not.

2 Materials and Methods

2.1 Study Area

As shown in Fig. 1, the study was carried out at Saharanpur city, Uttar Pradesh. The Saharanpur district is very near to the Shivalik hills range and lies under the upper Ganga-Yamuna region of northern India [10]. The mean sea level of Saharanpur district is ~269 m, and the annual mean rainfall is approximately 1150 mm [11]. Due to the deposition of alluvium soil across the district by the tributaries of two rivers, the soil is fertile. The population of the district is 3,464,228 out of which 69% lives in rural locations [12]. Thus, mostly the population depends on agriculture for their livelihood. The important industries in Saharanpur include the tobacco industry, cotton industry, paper mill, sugar industry and woodwork industry. The majority of the population depends on hand pumps and bore wells for water requirements. The samples for the study were collected from 18 different sites as shown in Fig. 2.

2.2 Sample Collection

Groundwater samples from hand pumps and bore wells of different locations were collected in bottles (polyethylene) which were prewashed by diluted acid and soaked with deionized water. Before sampling, hand pumps and bore wells were pumped for 10 min to remove standing water from the sources to get a representative sample. The samples were properly preserved and carried to the laboratory of Environmental

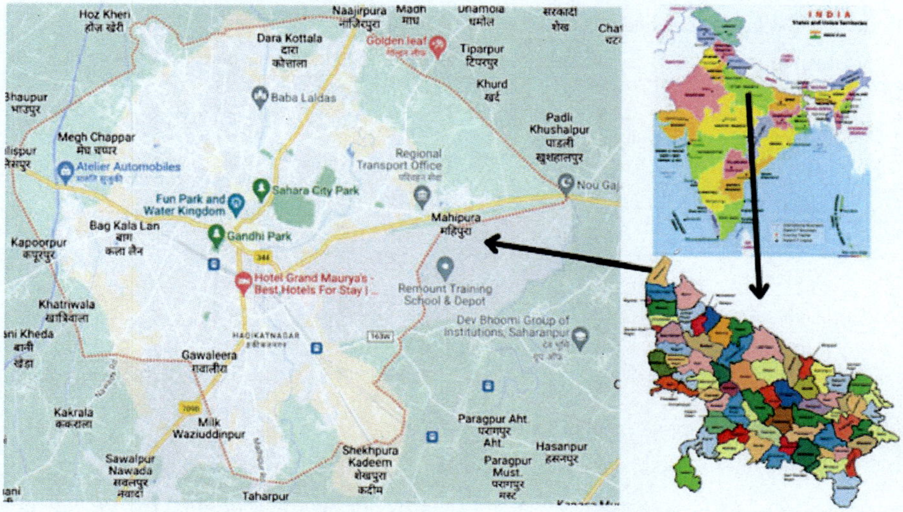

Fig. 1 Image showing study area

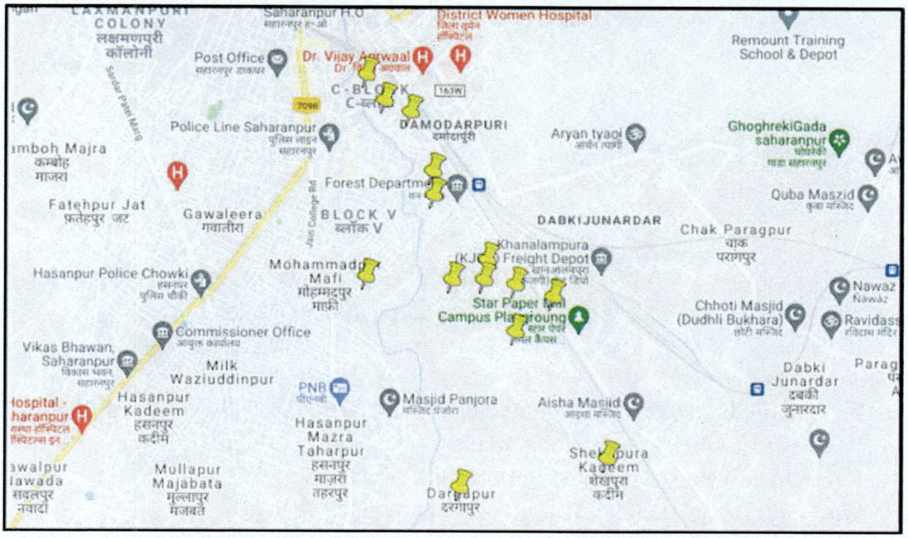

Fig. 2 Image showing sites of sample collected

Management Division, Central Pulp & Paper Research Institute, Saharanpur, India, for further analysis. The analyses were carried out as per standard [13]. All the samples were analyzed in duplicate to ensure more accuracy and less error.

In the present study, seven important parameters were chosen for the calculation of WQI. The standards for drinking water quality recommended by World Health Organization (WHO) [14], Bureau of Indian Standards (BIS) [15] and Indian Council for Medical Research (ICMR) were taken to compute Water Quality Index (WQI).

Table 1 Drinking water standards as per recommended agency [14, 15]

S. No	Parameter	Standard (Sn)	Recommended agency	Relative weight (Wn)
1	pH	6.5–8.5	ICMR/BIS	0.219
2	Electrical conductivity (EC)	300	ICMR	0.371
3	Total alkalinity	120	ICMR	0.0155
4	Total hardness	300	ICMR/BIS	0.0062
5	Chloride	250	ICMR	0.0074
6	Calcium	75	ICMR/BIS	0.025
7	Magnesium	30	ICMR/BIS	0.061
				$\sum$Wn = 0.7051

Note All values are in mg/l except pH and EC (μS/cm)

The weighted arithmetic index method [16] was adopted to calculate the WQI (Table 1)

$$WQI = \frac{\sum qnWn}{\sum Wn}$$

where

qn = Quality rating (nth water quality parameter) and n = 1, 2, ……0.6.

Wn = Relative weight of nth parameters.

Now,

$$qn = \frac{100(Vn - Vio)}{(Sn - Vio)}$$

where

Vn = Estimated value (nth parameter).

Sn = Permissible value (nth parameter).

Vio = Ideal value (nth parameter for pure water).

Vio = 7.0 (for pH) and, 0 (for all other parameters).

And,

$$Wn = \frac{K}{Sn}$$

where

K = Proportionality Constant.

$$\text{Now } \mathbf{K = 1/\sum (1/Sn)}$$

Table 2 Degree of water quality based on the value of WQI

WQI ranges	Degree of water quality
0–25	Excellent
26–50	Good
51–75	Poor
76–100	Very poor
>100	Unsuitable

The water quality index describes the quality of water as per Chatterji et al., 2002, given in Table 2.

3 Results and Discussion

The results of the physiochemical analysis of groundwater samples of different locations are presented in Tables 3 and 4 and Fig. 3.

The analysis results of different parameters of groundwater samples as presented in Table 3 reveal that only pH and Chloride concentrations meet the permissible limits as per the recommended agency. All other parameters were not meeting the prescribed standard permissible limits, except one or two locations for a specific parameter. The EC levels in all samples were found to be high in all locations. Only three sites were found to be ≤0.5 mS/cm, which was close to the permissible limit, i.e., 0.3 mS/cm. Out of 18 locations, only one location was found to meet the permissible limit of Total Alkalinity. For Total Hardness and Calcium, only 8 locations were found to meet the standard limit. Nine locations were found to meet the permissible of magnesium.

The possible impacts on groundwater quality may be likely due to the discharge of untreated sewage water and industries' effluent into river streams, as the city is an industrial hub of all kinds of large- and medium-scale industries. A seasonal river named Dhamola is also flowing on the side of the selected locations carrying municipal, household and industrial wastewater. The wastewater and waste are dumped into the river without any treatment. This may also degrade the groundwater quality by contaminating the groundwater aquifers through sediment percolation.

4 Conclusion

Among all the sampling locations, the value of different parameters varies significantly due to various anthropogenic means. Understanding the groundwater quality is important because it decides the suitability of water for different purposes like drinking, bathing, cooking, etc. It is difficult to understand the suitability of specific parameter results because all the parameters are not under permissible limits. Thus,

Table 3 Physiochemical analysis of different samples

Locations	pH	EC (mS/cm)	Total alkalinity (mg/l)	Total hardness (mg/l)	Chloride (mg/l)	Calcium (mg/l)	Magnesium (mg/l)
Himmatnagar	7.12	1.12	436	440	86.37	99	47
IPT campus	7.57	0.5	285	226	17	51	24
Indra Gandhi Colony-I	7.13	0.51	420	250	27.19	57	26
Indra Gandhi Colony-II	7.05	0.56	340	225	28.4	58	22
Near paper mill	7.01	0.93	450	357	50.18	79	39
Rajvihar	7	1.18	506	542	71	145	43
Anjani Vihar	7.39	0.75	352	352	58.39	82	29
Kapil Vihar-I	7.35	1.1	650	532	117	117	58
Kapil Vihar-II	7.25	0.93	458	384	56.38	88	39
Brahmpuri Colony	7.1	0.71	416	239	50.18	56	24
Brijesh nagar	7.6	0.7	398	225	95.17	58	22
Shastrinagar-I	7.54	0.96	512	408	79	112	31
Shastrinagar-II	7.42	0.9	470	435	70.4	102	44
Vinay vihar	7.54	0.64	400	294	37.38	68	30
Dargapur	7.29	1.01	588	388	69	95	37
Shekhpura-I	7.14	0.5	608	228	51	61	18
Shekhpura-II	7.18	1.71	614	500	151	146	33
CPPRI colony	8.24	0.45	55	210	8	48.1	22.1

WQI is formulated for the water with seven different water quality parameters to understand the quality of water in a single term. The study reveals that the WQI of all the locations was greater than 100. And in some locations, it was found to be more than 200. It means that the quality of water in these locations is extremely poor and not suitable for drinking purposes. The study provides useful information to plan and execute suitable practices to combat groundwater pollution in the study area.

Table 4 Calculation of water quality index (WQI) for different locations

Sample location	pH	qn (pH)	EC (mS)	qn (EC)	Calcium (mg/l)	qn (Ca)	Total hardness (mg/l)	qn (TH)	Chloride (mg/l)	qn (Chloride)	Total alkalinity (mg/L)	qn (TA)	Magnesium (mg/l)	qn (Mg)	WQI
Himmatnagar	7.12	24	1.12	373.33	99	132	440	146.67	86.37	34.548	436	363.33	47	156.67	231.76
IPT campus	7.57	114	0.5	166.67	51	68	226	75.333	17	6.8	285	237.5	24	80.00	138.39
Indra Gandhi Colony-I	7.13	26	0.51	170	57	76	250	83.333	27.19	10.876	420	350	26	86.67	116.26
Indra Gandhi Colony-II	7.05	10	0.56	186.67	58	77.333	225	75	28.4	11.36	340	283.33	22	73.33	117.42
Near paper mill	7.01	2	0.93	310	79	105.33	357	119	50.18	20.072	450	375	39	130.00	188.21
Rajvihar	7	0	1.18	393.33	145	193.33	542	180.67	71	28.4	506	421.67	43	143.33	237.37
Anjani Vihar	7.39	78	0.75	250	82	109.33	325	108.33	58.39	23.356	352	293.33	29	96.67	175.65
Kapil Vihar-I	7.35	70	1.1	366.67	117	156	532	177.33	117	46.8	650	541.67	58	193.33	250.88
Kapil Vihar-II	7.25	50	0.93	310	88	117.33	384	128	56.38	22.552	458	381.67	39	130.00	203.80
Brahmpuri Colony	7.1	20	0.71	236.67	56	74.667	239	79.667	50.18	20.072	416	346.67	24	80.00	148.84
Brijesh Nagar	7.6	120	0.7	233.33	58	77.333	225	75	95.17	38.068	398	331.67	22	73.33	177.48
Shastrinagar-I	7.54	108	0.96	320	112	149.33	408	136	79	31.6	512	426.67	31	103.33	227.06
Shastrinagar-II	7.42	84	0.9	300	102	136	435	145	70.4	28.16	470	391.67	44	146.67	211.63

(continued)

Table 4 (continued)

Sample location	pH	qn (pH)	EC (mS)	qn (EC)	Calcium (mg/l)	qn (Ca)	Total hardness (mg/l)	qn (TH)	Chloride (mg/l)	qn (Chloride)	Total alkalinity (mg/L)	qn (TA)	Magnesium (mg/l)	qn (Mg)	WQI
Vinay vihar	7.54	108	0.64	213.33	68	90.667	294	98	37.38	14.952	400	333.33	30	100.00	166.01
Dargapur	7.29	58	1.01	336.67	95	126.67	388	129.33	69	27.6	588	490	37	123.33	222.52
Shekhpura-I	7.14	28	0.5	166.67	61	81.333	228	76	51	20.4	608	506.67	18	60.00	116.49
Shekhpura-II	7.18	36	1.71	570	146	194.67	500	166.67	151	60.4	614	511.67	33	110.00	340.86
CPPRI colony	8.24	248	0.45	150	48.1	64.133	210	70	8	3.2	54.5	45.417	22.1	73.67	166.25

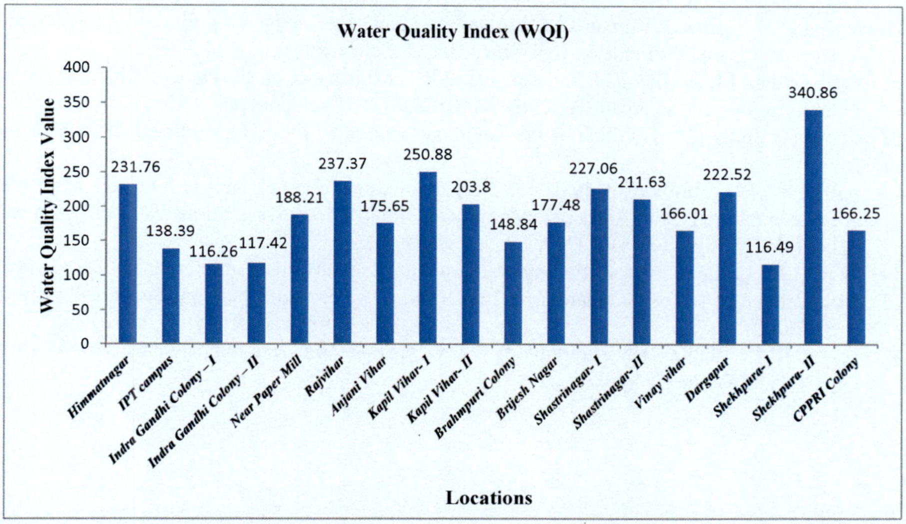

Fig. 3 Water quality index (WQI) for different locations

References

1. Majid SN, Khwakaram AI, Hama NY (2012) Determination of water quality Index (WQI) for Qalyasan stream in Sulaimani city/ Kurdistan region of Iraq. Int J Plant Anim Environ Sci ISSN 2231–4490
2. Kumar A, Malyan SK, Kumar SS, Dutt D, Kumar V (2019) An assessment of trace element contamination in groundwater aquifer of Saharanpur, Western Uttar Pradesh, India. Int J Elsevier Biocataly Agric Biotechnol 20:101213. https://doi.org/10.1016/j.bcab.2019.101213
3. Arivalagan P, Singaraj D, Haridass V, Kaliannan T (2014) Removal of cadmium from aqueous solution by batch studies using Bacillus cereus. Ecol Eng 71:728–735
4. Bhattacharjee S, Saha B, Saha B, Uddin MS, Panna CH, Bhattacharya P, Saha R (2019) Groundwater governance in Bangladesh: established practices and recent trends Ground. Sustain Dev Round Sustain Dev 8:69–81
5. Gupta DK, Bhatia A, Kumar A, Das TK, Jain N, Tomer R, Malyan SK, Fagodiya RK, Dubey R, Pathak H (2016) Mitigation of greenhouse gas emission from the rice-wheat system of the Indo-Gangetic Plains: through tillage, irrigation and fertilizer management. Agric Ecosyst Environ 230:1–9
6. Kumar A, Kumar V, Dhiman N, Ojha A, Bisen P, Singh A (2016) Consequences of environmental characteristic from livestock and domestic wastes in wetland disposal on groundwater quality in Lucknow (India). Int J Environ Res Public Health 3:112–119
7. Pugazhendhi A, Boovaragamoorthy GM, Ranganathan K, Naushad M, Kaliannan T (2018) New insight into effective biosorption of lead from aqueous solution using Ralstonia solanacearum: characterization and mechanism studies. J Clean Prod 174:1234–1239
8. Shanmugaprakash M, Venkatachalam S, Rajendran K, Pugazhendhi A (2018) Biosorptive removal of Zn (II) ions by Pongamia oil cake (Pongamia pinnata) in batch and fixed-bed column studies using response surface methodology and artificial neural network. J Environ Manag 227:216–228
9. Tyagi A, Mondal P, Siddique NA (2017) Quality assessment of ground water using water quality index at Yamunanagar, Haryana. In: Advances in Health and Environment Safety. Springer Transactions in Civil and Environmental Engineering, pp 369–376. https://doi.org/10.1007/978-981-10-7122-5_36

10. Malyan SK, Kumar J, Kumar SS (2014) Assessment of groundwater pollution of Saharanpur district, western Uttar Pradesh, India. Int J Recent Scient Res 5:1112–1115
11. IMD (Indian Meteorological Department) (2017) Ministry of earth sciences. Government of India. http://www.imd.gov.in/Welcome%20To%20IMD/Welcome.php
12. Census of India (2011) Office of the Registrar General of India. Government of India, New Delhi
13. APHA (2017) Standard methods for the examination of water and wastewater, 23rd edn. American Public Health Association/American Water Works Association/Water Environment Federation, Washington, DC, USA
14. WHO (2011) Guideline for drinking-water quality, 4 en. World Health Organization, Geneva
15. BIS (Bureau of Indian Standards) (2012) Drinking water—specification (Second Revision), ICS 13 060 20
16. Brown RM et al (1972) A water quality index-crashing the psychological barrier. Ind Environ Qual 1(1):173–178

Safety Concerns and Consequences of Cloud Seeding Implications—A Systematic Review

Akshi Kunwar Singh, B. Abhijith, and Leelakant Dewangan

Abstract The cloud seeding process is the most common weather modification technique carried out extensively since its invention in the 1950s. There are various benefits of cloud seeding that have helped many countries facing a severe loss of rain and scarcity of drinking water, but the repetitive and prolonged usage of the existing cloud seeding methods carries potential risks along with it that can be toxic, harmful and have effects on the metrology of a region. The risks associated with various chemicals are discussed along with the metrological data showing the harms caused and improvement in rainfall following the Cloud seeding method. These changes were deeply discussed, and the necessity for alternate seeding materials or alternate methods is focused on in this paper to consider the motto of environmental protection.

Keywords Cloud seeding · Static seeding · Dynamic seeding · Hygroscopic cloud seeding

1 Introduction

Various methods are developed to change the weather conditions in the environment. One of those methods is Cloud Seeding, by using chemicals such as AgI (Silver Iodide) to increase the precipitation rate, prevent hailstorms and reduce hurricanes [1]. The chemicals are dispersed in the cloud to increase the rainfall and snowfall in a particular area, which initiates the formation of the ice particles as the chemical crystals act as a nucleus [2]. When the size of these formed particles is such that they cannot remain suspended, then the particles start to fall due to gravity [3]. If this happens at higher altitudes or cold regions, such as mountains, the particles may fall as snow, but the falling ice particles in the lower altitude region convert into water droplets and will fall as rain [4].

A. K. Singh (✉) · B. Abhijith · L. Dewangan
School of Engineering, UPES, Sustainability Cluster, Dehradun, India
e-mail: singh.akshi85@gmail.com

N. A. Siddiqui et al. (eds.), *Advances in Waste Management*, Lecture Notes in Civil Engineering 301, https://doi.org/10.1007/978-981-19-7506-6_2

Many countries like the UAE, the USA and China have adopted the cloud seeding process. It is an age-old method used to increase the precipitation rate by using a chemical process to modify the weather. Chemicals [(silver Iodide or dry ice, Sodium Chloride and calcium carbide) = Ice nuclei] are sprayed into already-formed clouds to increase the precipitation rate [5]. The biggest drawback of the cloud seeding method is that this method can only be used when clouds are present because no new clouds are formed. An average of 10–15% increase in precipitation is observed after cloud seeding [6]. Even though many studies show an increase in the water or precipitation, the increase in precipitation is minimal in some cases. This increase in precipitation may happen due to many factors, such as a change in the average weather pattern.

Various advanced studies have been done to interpret the data associated with cloud seeding and its effectiveness; in addition to it, microscopic studies were also conducted to understand the adverse effects of these seeding processes in the longer run. The studies done in India are of minimal importance than in any other region as IIT Madras and Rain and Cloud Physics Research centre (RCPR) conducted most of their studies in the 90s. However, the current trend is deteriorating, and the focus on creating less toxic and more efficient alternatives is minimal. As one can see, most of the chemicals used since the 1970s are still being used without much variance. This paper discusses various possessing threats by the overused chemical, resulting in adverse climatic and toxic conditions in the environment.

1.1 How Cloud Seeding Works

A specified seeding agent is added to the particular type of cloud (Cumulonimbus: very few types of this cloud result in an increase in the precipitation rate) as shown in Fig. 1. The spraying of chemicals can be done by various methods, such as using a plane and spraying the chemicals on the top layer of the cloud. In another method, rockets are used to spread the chemical by creating explosions near the clouds. In ground generators, acetone-containing AgI is burned to produce the ice crystal, which goes up with smoke. The chemicals used in the cloud seeding method also come to the ground with rain and accumulate in the environment if the seeding is performed many times [4]. The chemical used for the seeding purpose has the same crystal structure as ice [7]. Silver Iodide is released in the hygroscopic method into the atmosphere from the plane. Silver iodide is the most preferred chemical because of glaciogenic (Ice Nucleating) and has similar properties to ice crystals [8]. Thus, it acts as an ice nucleus that initiates precipitation in the form of ice particles. When this ice gets heavy, it falls as rain or snowfall due to gravity. If cloud seeding is performed in the mountain region, higher altitude region or the colder region, the precipitate falls as snow. And if it is done in a low-level region or a warmer region, ice transforms into water droplets, and rainfall occurs.

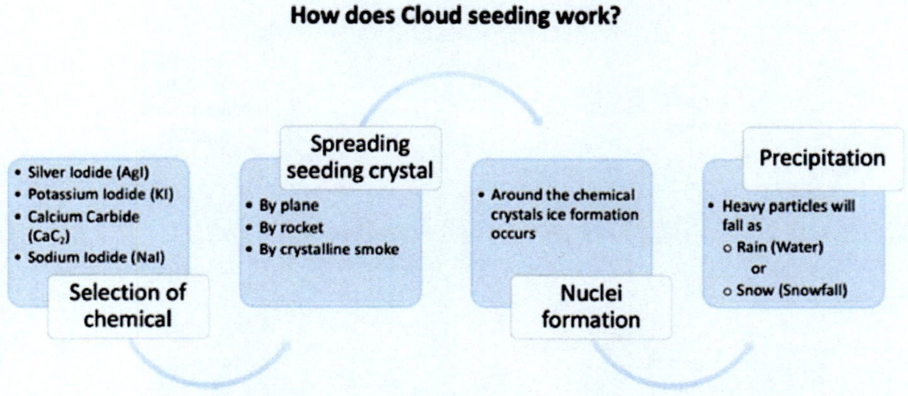

Fig. 1 Steps for cloud seeding

Silver Iodide (Ice nuclei) + Light tiny rainwater droplet (Already present in the cloud) = Ice Particles (Heavy enough to fall).

The clouds do not live long, but particles collide on their own and turn into larger, heavier particles, so every fraction of particles of a raindrop is sufficient to convert into a larger one.

1.2 Important Factors in Cloud Seeding Operation

As shown in Fig. 2, various factors affect cloud seeding effectiveness, as choosing the correct type of cloud is very important because not all clouds start to precipitate when needed. Cumulonimbus is one of the clouds which gives heavy rain when seeded. Also, the seeding time is significant because if the cloud is seeded early or later, the effectiveness of the cloud seeding will be significantly less. Another factor is the amount of the seeding material used, which should be enough to initiate and ensure that further nucleation is self-sustained, whereas another method used for cloud seeding involves proper and homogenous dispersion of the chemical on the cloud, which increases the efficiency of the cloud seeding process [9].

The correct type of material selection is significant, and it plays a vital role in cloud seeding because the size, structure and properties of these materials affect the effectiveness of cloud seeding and can alter the precipitation [10]. Some essential chemicals which are used in the seeding process are silver iodide (AgI), liquified propane (C_3H_8), which, when burnt, produces smoke containing ice nuclei, solid CO_2, sodium iodide (NaI), calcium carbide (CaC_2) and calcium chloride ($CaCl_2$) [11]. The most commonly used chemical is AgI due to its crystal structure similarity with ice. Environmental parameters such as pressure, temperature, humidity, speed, the direction of airflow and nuclei present in the atmosphere manipulate the seeding effectiveness. The environmental parameters also contain local natural structures

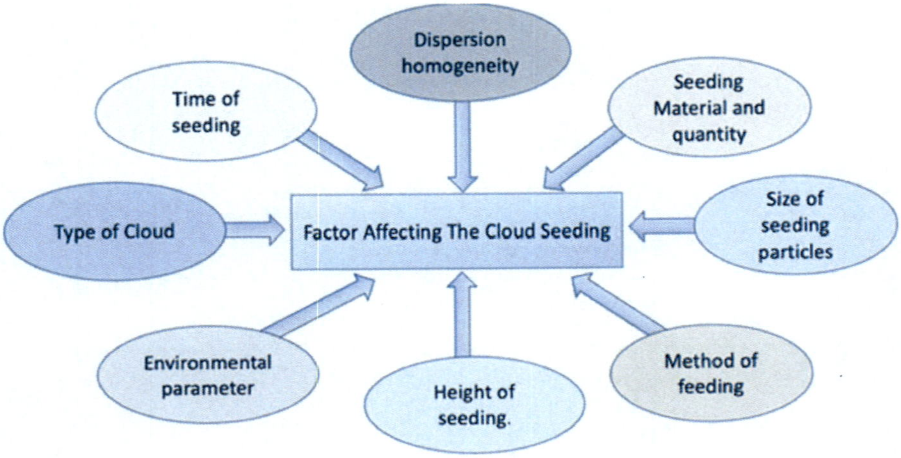

Fig. 2 Factors affecting cloud seeding

such as soil conditions, plants and mountains. Usually, the effect on cloud seeding due to temperature change is minimal and does not lead to much change in the effectiveness of cloud seeding. However, when all the parameters change, it causes a significant variation in the effectiveness.

Along with these factors that affect the clouds while performing the cloud seeding operation, some factors affect the effectiveness of the cloud seeding by altering the nucleation. This altering of nucleation can be done by changing the seeding methodology [10]. So, all the parameters must be controlled or appropriately checked before performing the cloud seeding operation.

2 History of Cloud Seeding

Throughout the world, myths were followed before the scientific studies of why rain occurs. Certain people believed that firing artillery into the sky could result in rain, including blood sacrifice [12]. So, to make rain happen, various parts of the world followed various pseudo-scientific methods, including using frogs by hanging them on trees, making giant paper dragons hoping for rain and in certain parts, humans were also sacrificed for rain. Figure 3 depicts the history of cloud seeding years.

These myths were cleared when James Pollard Espy, who is considered the father of the US Weather Bureau, explained the method of cloud formation [13]. He proved the science behind the theory of fire producing rain and explained it with the theory that hot air expands, resulting in condensing of air vapours and a drop in temperature. He proposed that burning a large amount of land could result in rain. Cloud seeding experiments were first conducted in 1946 by Vincent J Schaefer, an American metrologist and chemist [13–15].

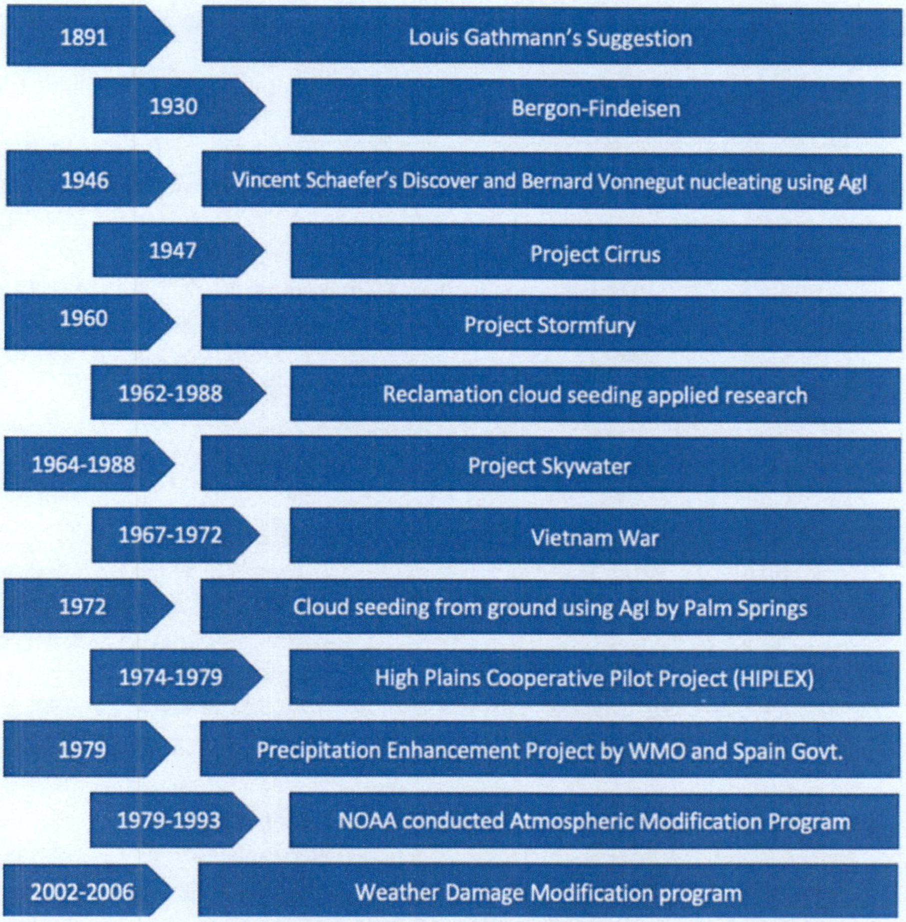

Fig. 3 Timeline of history of cloud seeding

2.1 Cloud Seeding in UAE

Even though the cloud seeding studies were conducted in the 1940s, the UAE started these studies in the 1990s. Even though the start was late, the UAE was never lagging behind the technology; in 2005, the UAE launched various awards and rewards for studies and innovations on weather modification techniques and technology [16]. In 2010, cloud seeding started as a project for altering the weather across Dubai and Abu Dhabi deserts. In 2015, UAEREP (UAE Research Program for Rain Enhancement Science), the programme for rain enhancement science, allowed scientists and researchers to indulge in this field. By 2014, a total of 187 cloud seeding missions were conducted in the UAE alone and were increased to 214 missions in 2017, with a total of 247 missions in 2019 for altering the weather with the help of technology artificially [17]. Figures 4, 5 and 6 depict the average rainfall in Sharjah with or without cloud seeding, and their comparison in mm precipitated.

The comparison of data with and without cloud seeding in the UAE indicates the importance of cloud seeding, as we can see that from 1992 to 2007, the rainfall average was 13.68 mm without cloud seeding, which is a minimal value considering the rainfall and indicates a very short rainfall bringing less amount of water to the area

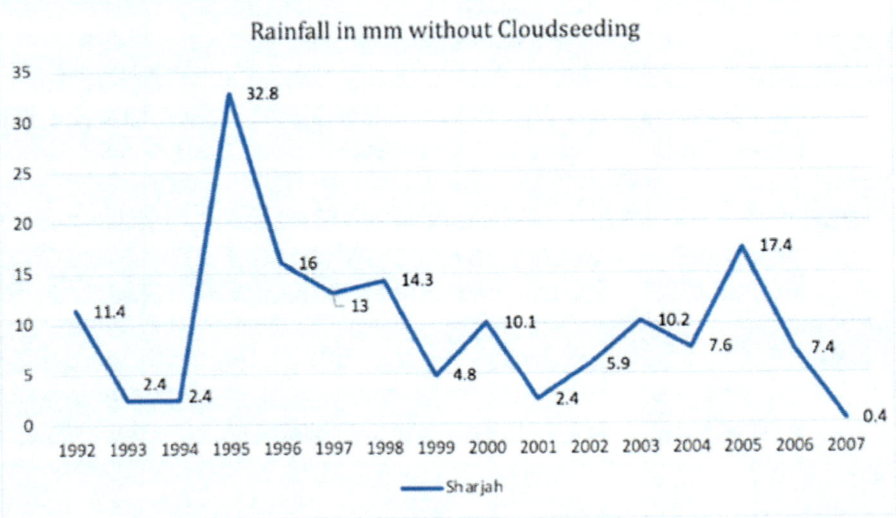

Fig. 4 Rainfall record of Sharjah in mm without cloud seeding [18]

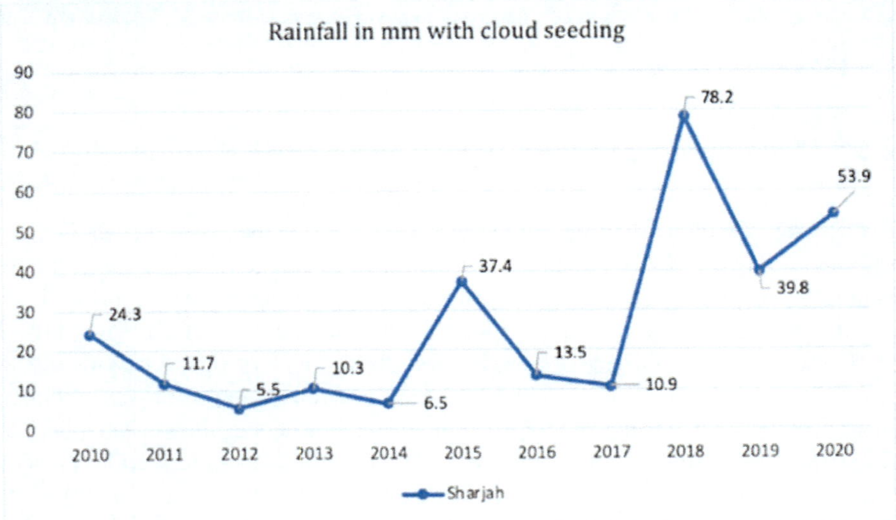

Fig. 5 Rainfall record of Sharjah in mm with cloud seeding [18]

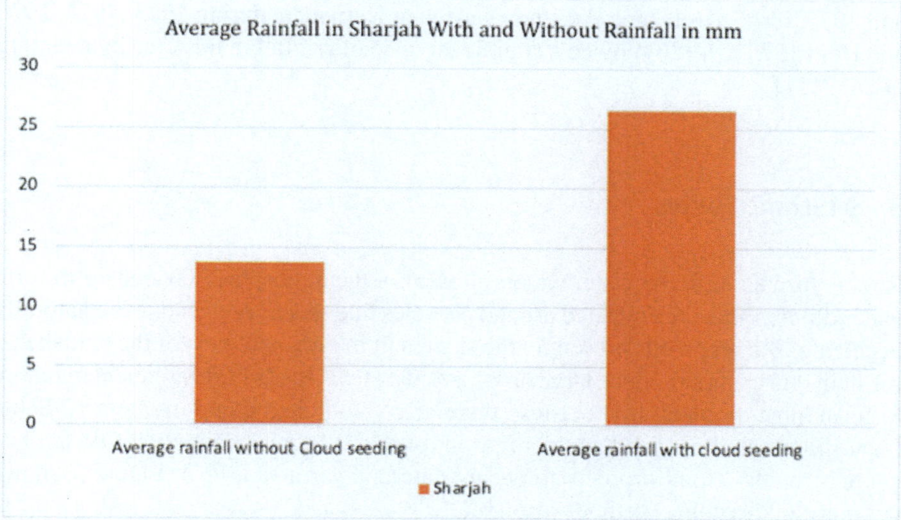

Fig. 6 Average rainfall comparison with and without cloud seeding [18]

[16]. Since the project of cloud seeding, as mentioned in 2010, the rainfall increased by a significant value bringing medium to heavy rainfall with an average of 26.55 mm of rainfall which increased by 194% [17]. This value was nearly double that without cloud seeding, and in specific years, the rainfall rose to 78.2 mm, indicating heavy rainfall leading to flooding [18].

2.2 Cloud Seeding in India

In India, rainfall is becoming scarce in certain states, and the decline in rain has increased drought in various states. Regularly, it is necessary to implement methods to overcome this drought for drinking water, agriculture and other purposes as most of the Indian market is focused on the agricultural sector, and droughts could impact it adversely. In 1951, the Tata firm used AgI for cloud seeding in the Western Ghats using ground generators. Later in 1952, Dr Banerji tried using salt-based seeding materials and silver iodide using a balloon released from the ground filled with hydrogen gas, allowing it to reach sufficient altitude [19]. In 1953, Council of Scientific and Industrial Research initiated the setting up of Rain and Cloud Physics Research (RCPR) for having an extensive and deep study on conducting long-term seeding programmes. With the use of salt-based seeding material, the Northern part of India experienced an increase in rainfall by about 20% from 1957 to 1966. RCPR integrated with IIT madras and conducted various cloud seeding studies till 1977. Various experiments were carried out in Uttar Pradesh and Karnataka during 1974 and 1975, respectively, and the study carried at Karnataka during 1973, 1974, 1976 and from 1979 to 1986 showed a significant raise of rainfall in this area by a margin of 24% [11].

3 Methodologies

Rain is formed after the water vapour present in the atmosphere condenses to form water droplets that are collected around particles like specks of dust that are naturally occurring. But all clouds present in the sky result in rain, and most of the clouds that result in rain cannot produce large drops as only a few clouds have sufficient moisture to form huge droplets. It is because there aren't sufficient cloud droplets available for forming raindrops, and another reason for this is because rain clouds do not last enough for the cloud droplets to accumulate and form rain [20]. Figure 7 shows various cloud seeding methods available.

Cloud seeding is the process that helps in increasing the number of ice crystals in the cloud as they provide additional nuclei where the water droplets can accumulate through condensing, as shown in Fig. 8. These nuclei can be formed with the help of materials like salts, silver iodide, potassium iodide and dry ice, and in particular studies, liquid propane is also used for this [21]. Liquid propane can produce ice crystals at a more significant temperature than silver iodide; hence, the applicability is also greater [22]. In expert studies in this field with silver iodide to induce rain, it was found that the AgI have a structure that is identical to that of natural ice crystal structure, and on applying the AgI in the topmost part of a cloud that is on developing or growing stage, the AgI grows faster than the naturally forming ice crystals making the crystal heavy due to the accumulation of moisture leading to the formation of large raindrops. Studies have identified that just a single gram of AgI can lead to 10 trillion artificial crystals of ice [16].

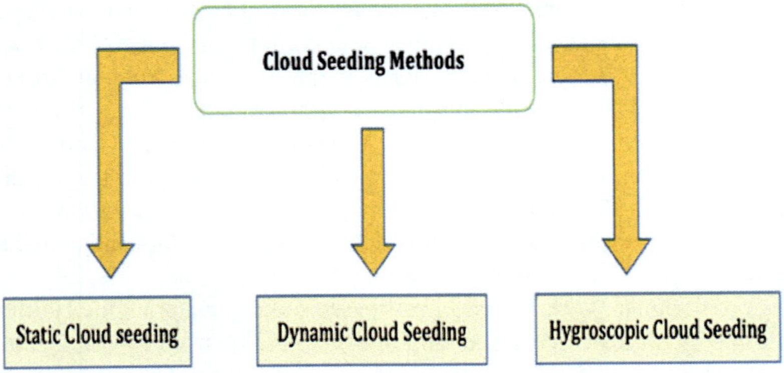

Fig. 7 Cloud seeding methods

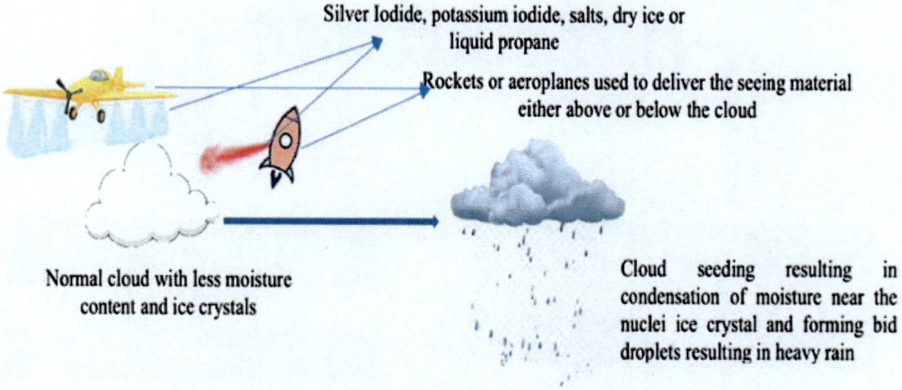

Fig. 8 Cloud seeding process

3.1 *Static Cloud Seeding*

The static cloud seeding technique is the method of spraying Silver Iodide or any other potential seeding material into the clouds. It helps the crystal formation in the cloud, which accumulates the nearby moisture to rapidly cool by accelerating the speed of dispensing the water, thus reducing the time of cloud formation to rain. For the statistical seeding process, the window for successful seeding is limited. If successful, it causes the alteration in the microstructure of the cloud, increasing the concentration of ice crystal, rapid development of precipitation and reducing the liquid water content that is supercooled [23]. Due to the limited window of success for static seeding experiments, the results vary from successful seeding to a reduction in rainfall or no effect on seeded clouds [19]. Hence, the adoption of this method can never be evidenced as successful in every region unless all environmental conditions match that of the successful seedings. Static cloud seeding doesn't require air current, and the condition during seeding will be primarily stable [11].

Static cloud seeding hence is limited to the following cloud conditions:

- This method has the potential to be successful for cold or continental clouds.
- Clouds with temperatures varyg in the maximum range of $-10°$ to -25 °C.
- The timing should be managed such that this process needs to be done before the depletion of supercooled water through precipitation or entrainment [20].

3.2 *Dynamic Cloud Seeding*

The primary focus of this method is to enhance the vertical airflow ensuring the vertical accumulation of more water resulting in more precipitation. The significant difference between dynamic and static seeding is the role of air current, and the dynamic seedings objective is to utilise and boost the vertical air current, allowing

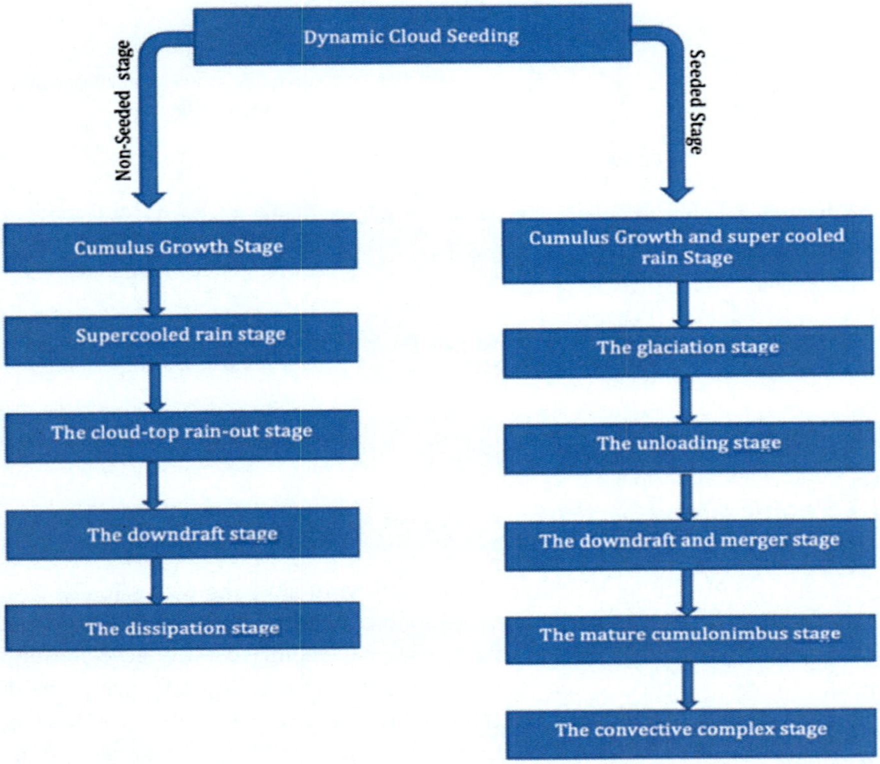

Fig. 9 Dynamic cloud seeding stages

more water to pass through the clouds, resulting in more water condensation, resulting in more rain [21]. The quantity of seeded particles induced into the clouds is also much more incredible considering the static seeding process where per litre of water, the objective is to ensure 100–1000 ice crystals, and this number corresponds to the requirement of dropping 200–1000 g of silver iodide through flares to the highly condensed supercooled liquid water [18]. This cloud seeding method observed glaciating evens in the cloud, converting the cloud from liquid to ice. This initiative is being done in Thailand and Texas; the results derived are satisfactory compared to statistics where the window is limited in case of success. Figure 9 depicts the cloud seeding stages.

3.3 Hygroscopic Seeding

The formation of precipitate in warm clouds occurs due to the frequent collision of the clouds, and due to the effectiveness of this process, the clouds contain sufficient substantial water content in liquid form. Improving the precipitation amount in the

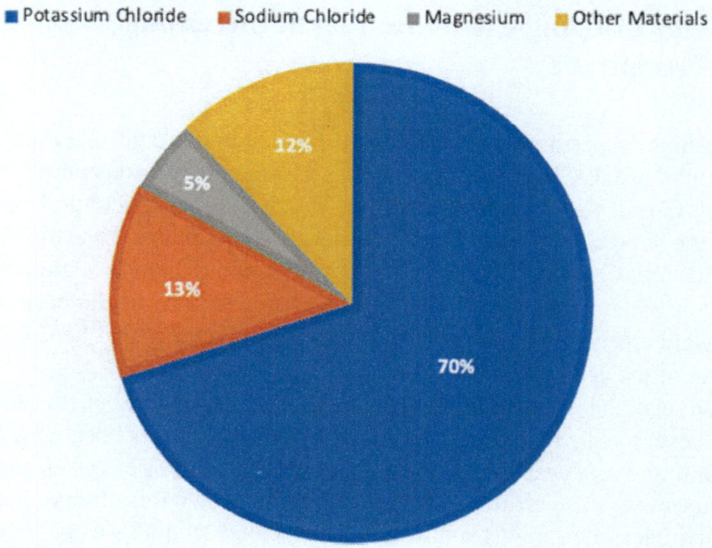

Fig. 10 Materials used in hygroscopic cloud seeding [16]

cloud is by inducing hygroscopic salts, which absorb water through vapour deposition [18]. This method involves the deposition of small salt particles deposited in the base of the clouds, which on vapour deposition absorbs moisture and increases the salt's size. Once more prominent, they form one mass embryo that can participate in a collision for rain formation.

The term hygroscopic means the capability of a substance to intrigue moisture from the air. This method is applied in warm clouds using materials like sodium, potassium and magnesium salts, and these materials are considered the embryo around which the raindrop forms [16]. This type of seeding process is mainly used in the UAE, where the presence of warm clouds is more.

Applicability of the materials mentioned in Fig. 10; Potassium Chloride consisting of 70% and Sodium Chloride consisting of 13% of the flare composition helps attract the water droplet. Magnesium composing 5% of the flare composition controls the flame from the flare. The other material consisting of 12% is solidarity material that helps in binding these other materials together.

3.4 Ground-Based Cloud Seeding

In ground generators, acetone containing AgI is burned to produce the ice crystal, which goes up with smoke. The chemicals used in the cloud seeding method also come to the ground with the rain and accumulate in the environment if the seeding is performed many times [4].

4 Benefits of Using Cloud Seeding in Increasing the Precipitation

Cloud seeding is performed to increase the precipitation and make more rain or snowfall than usual. Cloud seeding can be performed for various purposes, such as increasing rain in the drought area so that it can help in agricultural purposes, to increase the snowfall to maintain the ice level of mountains, to avoid flooding in flood-prone areas by identifying the clouds and seeding them to some other places, and also to avoid rain or snowfall in a particular area for a particular occasion. Cloud seeding can be used for agriculture to increase productivity and hence can lead to the growth of the economy [11].

If cloud seeding is used to suppress the hailstorm, the per acre benefit can be $14.65 to $ 23.35 every year. This value cumulatively for North Dakota becomes from $300 to $477 million every year [24]. In the winter Olympics, China has used cloud seeding to make the event successful. In China, decrease in snowfall is observed, and for the winter Olympics, the snow is significant. Thus, cloud seeding was performed in the Haituo mountains to make snowfall [25].

5 The Potential Risk Possessed by Cloud Seeding

5.1 Risk from AgI

Right now, similar to ice crystalline structure, AgI is the most commonly used chemical, which is not commonly available in the atmosphere. And the concentration of free silver in the soils and sediments due to the cloud seeding is deficient. So, it does not pose any toxicological threat. But the accumulation of the AgI in the same place due to the repeated seeding process makes the concentration moderately toxic.

In the laboratory, a simulation of the environmental concentration of 12.5 micromolar of the AgI after repeated seeding identified its adverse effects on the various microbes in soil and water. The highest increase of about 26% in demand for BOD_5 was 2.5 micromolar. But soil worms were not affected by the AgI. There is a considerable drop in the photosynthesis of the producers [4] (Tables 1 and 2).

Table 1 Change in biological oxygen demand for different concentrations of Silver Iodide [4]

S. No	The concentration of AgI (micromolar)	Increase in the BOD_5 (%)
1	0.43	22.97 ± 6.37
2	2.5	26.57 ± 9.97
3	12.5	25.02 ± 6.66

Table 2 Inhibition of different concentrations of Silver Iodide [4]

S. No	The concentration of AgI (micromolar)	Inhibition
1	0.43	No effect
2	2.5	36.04 ± 75.78%
3	5–12.5	Significant reduction in luminescence

5.1.1 Soil Bacteria Viability

At 0.43 micromolar concentration, there is an increase in the number of dead cells in the taken microbes, leading to improper function of metabolism and causing a decreased population of soil bacteria [4].

5.1.2 AgI Test on Water

There is a considerable drop in the physical activity of the microorganisms at the concentration of 0.43 micromolar also. After some intervals, the numbers also decreased drastically, which shows the toxicity of the AgI. In the producers, decrease in the photosynthesis process was significant. Silver Iodide is a substance that comes under a hazardous category, and it is a primary toxic pollutant. Also, it is categorised as a non-soluble inorganic substance. In Australia, the Threshold Limit Value (TLV) value of AgI in drinking water is 0.43 micromolar. Free silver ion is the majorly identified toxic in the AgI. Free silver ions will be in the solid form, and this silver form is used in cloud seeding because it acts as nuclei in the cloud seeding process [4].

5.2 Impact of Potassium Iodide

Potassium iodide (KI) causes a significant increase in the iodine level in the body. 3 mg of KI per day for a 4-week female can severely increase the thyroid organs and cause a weight change. A small KI intake can affect thyroid hormones and hormone supply (THS) levels. Frequent intake of KI causes a considerable decrease in the sodium iodide symporter (NIS) mRNA and pendrin (PDS) mRNA, which can cause mutation of the species, and this mutation has some adverse effects on the body of the mammal [26]. These changes in the mother's thyroid gland behaviour can cause a severe impact on the body and brain growth of the foetus and the infant child because for the growth, they are dependent on the mother. These can even impair the brain of the foetus or infant [27]. However, the WHO recommendation for KI in adults is 130 mg, but for mothers with foetuses and infant child it is considerably less, and it is under another section type of population [28]. Foetus and an infant child are more in danger of getting severe effects of iodine [27].

5.3 Risk Posed by Sodium Iodide (NaI)

Frequent sodium chloride intake can cause a considerable decrease in the NIS mRNA and PDS mRNA, leading to the mammals' mutation [26]. In adults, iodide causes an increase in blood pressure and can lead to heart problems and thus may have a critical negative impact on body function. After reacting with the NO2, Iodide forms NO, which is acidic [29]. That can make the water acidic and can cause damage to plants and microorganisms. Also, if this happens inside the stomach, it can cause severe irritation.

5.4 Risk Posed by Calcium Carbide (CaC_2)

CaC_2 always has some amount of arsenic which is highly carcinogenic. Also, it contains a good amount of toxic material, which can reach the ground with rain or snowfall [30]. This condition can be very harmful not only to aquatic life and plant but also it could cause harm to humans too. The impact of CaC_2 is chronic, and it could have adverse effects in the long term. Some short-term effects are skin burns, irritation and inflammation [30]. Accumulation of calcium carbide in the soil can result in the concentration of arsenic in agricultural products such as rice [31]. When reached into the body, it can cause cancer in the long term.

5.5 Potential Risk of Flooding

Seeding on the Cumulonimbus clouds can lead to very severe flooding in the area of the Cumulonimbus cloud cover. The rainfall may happen at such a rate that it won't provide sufficient time to react to the people [11]. A scenario, similar to that stated, happened in South Dakota on June 9 and 10 in 1972 [32]. Flooding can cause heavy property damage and may damage agricultural land or other land areas and the cost of damage. Thus, flooding can impact the whole area for a very long time with a high cost, as that happened in the case of the Yuba city flooding in 1955, which was about $200,000,00 at that time, in 1955 [33]. It may have many fatalities leading to trauma in people's minds. Damage to dams and check dams can make this impact catastrophic. The trial lake dam break happened due to a cloud seeding operation on June 7, 1986 [34]. It also severely hampers other animals and organisms' human and natural habitats. Flooding causes the displacement and migration of a vast population.

5.6 *Potential Risk of Draught*

The clouds heading to the places where they generally pour their water or snow may face the drought condition if seeding is done before reaching the place. It changes the rain pattern and can have a severe effect in the long run if cloud seeding is practised [4]. The severity may lead to absolute draught or infertile land area, thus soil erosion. Using cloud seeding only in the flooding-prone cloud before reaching the area may be one of the solutions [35]. But the cloud pattern is not consistent every year, and it changes. Draught also forces the area's population to migrate and destroys the local natural habitat of the animals and other organisms. It also causes substantial economic crises and a food shortage in the area. Drought can cause an increase in the temperature, thus changing the natural pattern of wind and rain.

6 Conclusion

Researchers across the globe have done various studies to improve the nuclei ice crystallisation so that more precipitates can be collected to form huge clouds with heavy drops of water, resulting in receiving more mm of rainfall in the areas where they are performed. Cloud seeding has potentially shown its advantage in various countries suffering from fewer rains and more drought seasons, resulting in less agricultural loss and a lack of drinking water. Cloud seeding can be employed on other special occasions to facilitate the change in weather in the required time to avoid these challenging weather conditions.

However, there are many advantages of cloud seeding, but the study shows that the continuous use of these chemicals hurts the atmosphere, land and water bodies. Over-exhausting cloud-seeding resources in an area can decrease that region's precipitation level, causing no cloud formation and moisture accumulation that could cause drought for a more extended period, affecting the entire climatic statistics of the area. Also, the improper planning and application of the cloud seeding processes on clouds that are not within the safe temperature limits could lead to the formation of hailstorm clouds resulting in a heavy downpour causing flooding. Also, if the cloud seeding is carried out at locations which are improper or inaccurate can cause a change in an extreme weather pattern that could affect the livelihood, agriculture sector and economy of the location resulting in significant damage to property and life.

Even though there are many adverse effects of cloud seeding, the potential usage of cloud seeding is very high because of its applicability and merits for the receiving end. So further research is required to eliminate the toxicological effects of the chemical in the longer run. It is necessary to use scientific data, including satellite information and radar data, to apply the cloud seeding at the right time, place and quantity to avoid overexploitation, thus preventing flooding or drought. Also, studies should be done to find substitute seeding materials or compounds which are economical, non-toxic and sustainable to reduce the impact it causes on the atmosphere and nature.

References

1. Bailey E (2020) The hurricane minimizer. In: Conf. Proc.–IEEE SOUTHEASTCON, vol. 2. https://doi.org/10.1109/SoutheastCon44009.2020.9368294
2. Ćurić M, Janc D, Vučković V (2006) Cloud seeding impact on precipitation as revealed by cloud-resolving mesoscale model. Meteorol Atmos Phys 95(3):179–193. https://doi.org/10.1007/S00703-006-0202-Y
3. Thompson ARD (1986) Cloud Seeding Potential over Semi-Arid Lands. Water deficiencies in semi- arid regions have led to the development of a wide range of s, vol. 15(5), pp. 276–281. Springer on behalf of Royal Swedish Academy of Sciences Stable. https://www.jstor.org/stable/4313276
4. Fajardo C et al (2016) Potential risk of acute toxicity induced by AgI cloud seeding on soil and freshwater biota. Ecotoxicol Environ Saf 133:433–441. https://doi.org/10.1016/j.ecoenv.2016.06.028
5. Zheng W, Xue F, Zhang M, Wu Q, Yang Z, Ma S, Yu K (2020) Charged particle (negative ion)-based cloud seeding and rain enhancement trial design
6. Pokharel B, et al (2021) A modeling examination of cloud seeding conditions under the warmer climate in Utah, USA. Atmos Res 248:105239. https://doi.org/10.1016/j.atmosres.2020.105239
7. Rangno AL, Hobbs PV (1995) A new look at the Israeli cloud seeding experiments. J Appl Meteorol 34(5):1169–1193. https://doi.org/10.1175/1520-0450(1995)034%3c1169:ANLATI%3e2.0.CO;2
8. Yuan J, et al (2021) Cloud seeding evidenced by coherent doppler wind lidar. Remote Sens 13(19). https://doi.org/10.3390/rs13193815
9. Mather GK, Terblanche DE, Steffens FE, Fletcher L (1997) Results of the South African cloud-seeding experiments using hygroscopic flares. J Appl Meteorol 36(11):1433–1447. https://doi.org/10.1175/1520-0450(1997)036%3c1433:ROTSAC%3e2.0.CO;2
10. Vi PWD (1981) "'p~vi~w~d","
11. Malik S (2018) Cloud seeding; its prospects and concerns in the modern world–a review. Int J Pure Appl Biosci 6(5):791–796. https://doi.org/10.18782/2320-7051.6824
12. Lan D (1985) Guns and Rain: guerillas and spirit mediums in Zimbabwe, no. 38. University of California Press
13. Espy JP (1841) The philosophy of storms. CC Little And J. Brown
14. Schaefer VJ (1949) The formation of ice crystals in the laboratory and the atmosphere. Chem Rev 44(2):291–320
15. Schaefer VJ (1952) Formation of ice crystals in ordinary and nuclei-free air. Ind Eng Chem 44(6):1300–1304
16. Almansoori H, Badran A (2020) Cloud Seeding In The UAE Research Paper
17. Al Hosari T, et al (2021) The UAE cloud seeding program: a statistical and physical evaluation. Atmosphere (Basel) 12(8). https://doi.org/10.3390/atmos12081013
18. Almheiri KB, Rustum R, Wright G, Adeloye AJ (2021) Study of impact of cloud-seeding on intensity-duration-frequency (IDF) curves of Sharjah city, the United Arab Emirates. Water (Switzerland) 13(23):1–13. https://doi.org/10.3390/w13233363
19. Dash B (2019) Rain enhancement technology: making sense of the 'cloud seeding' program in India. Bull Sci Technol Soc 39(3–4):33–42. https://doi.org/10.1177/0270467620963708
20. Dennis AS (1980) Weather modification by cloud seeding. Academic Press
21. Girardo RB (1973) Project skywater. Agric Eng 54(3):10–13
22. Howell WE (1977) Environmental impacts of precipitation management: results and inferences from project skywater. Bull Am Meteorol Soc 58(6):488–501. https://doi.org/10.1175/1520-0477(1977)058%3c0488:eiopmr%3e2.0.co;2
23. Laaksonen A, Malila J (2022) Ice nucleation. In: Nucleation of Water, pp. 209–248. https://doi.org/10.1016/B978-0-12-814321-6.00018-X
24. Murphy P (1989) Economic impacts. Outdoor Recreat Canada 791:163–197. https://doi.org/10.4337/9781781002865.00019

25. Ma X, Bi K, Chen Y, Chen Y, Cheng Z (2017) Characteristics of winter clouds and precipitation over the mountains of Northern Beijing. Adv Meteorol 2017. https://doi.org/10.1155/2017/3536107
26. Lebsir D et al (2018) Effects of repeated potassium iodide administration on genes involved in synthesis and secretion of thyroid hormone in adult male rat. Mol Cell Endocrinol 474:119–126. https://doi.org/10.1016/j.mce.2018.02.017
27. Lebsir D et al (2019) Repeated potassium iodide exposure during pregnancy impairs progeny's brain development. Neuroscience 406:606–616. https://doi.org/10.1016/j.neuroscience.2019.02.016
28. WHO (WHO) (2017) Iodine thyroid blocking. In: Guideline use plan responding to Radiol. Nucl. emergencies. Geneva World Healing Organisation
29. Bailey SJ et al (2017) Influence of iodide ingestion on nitrate metabolism and blood pressure following short-term dietary nitrate supplementation in healthy normotensive adults. Nitric Oxide Biol Chem 63:13–20. https://doi.org/10.1016/j.niox.2016.12.008
30. Okeke ES, Okagu IU, Okoye CO, Ezeorba TPC (2022) The use of calcium carbide in food and fruit ripening: Potential mechanisms of toxicity to humans and future prospects. Toxicology 153112
31. Liu G, Meng J, Huang Y, Dai Z, Tang C, Xu J (2020) Effects of carbide slag, lodestone and biochar on the immobilisation, plant uptake and translocation of As and Cd in a contaminated paddy soil. Environ Pollut 266:115194. https://doi.org/10.1016/J.ENVPOL.2020.115194
32. Dennis A (2010) Cloud seeding and the rapid city flood of 1972 Rapid City, South Dakota. J Weather Modif 124–126
33. Mann DE (1968) the Yuba City flood: a case study of weather modification litigation. Bull Am Meteorol Soc 49(7):690–714. https://doi.org/10.1175/1520-0477-49.7.690
34. Julander R, Clayton J, Griffith D, Hasenyager C (2018) A proposal for a change in winter snowpack suspension criteria used in the conduct of cloud seeding programs in Utah. In: 86th Annu. West. Snow Conf., pp 53–62
35. Herschy RW et al (1998) Clouds (cloud seeding). Encycl Hydrol Water Resour 143–144

Review of Study on Various Forest Fire Detection Techniques Using IoT and Sensor Networks

Akshi Kunwar Singh, Sinan Mohamed Rafeek, P. S. Harikrishnan, and Irene Wilson

Abstract Forest is one of the essential pillars of the ecosystem. Moreover, it provides all the essentials required for our very own existence. However, forest fires have been a significant area of concern over the years. For this purpose, various devices have been used over the years. Internet of Things (IoT) has been a significant advancement in early detection and mitigation. IoT, along with various software, enables error-free data in real time. This paper has highlighted three significant advancements using IoT, enabling the early detection of a forest fire. Firstly, Convolution Neural Network (CNN) is used for forest fire detection. Another method utilises a series of sensor network paradigms to get real-time data for forest fires. The final method utilises an EP32 board along with sound, rain, DHT11, and PIR sensors for its swift and systematic reporting of forest fires.

Keywords Forest fire · Convolution neural network · MATLAB · Sensors · Satellite · Neural network · EP32 board · DHT11 · PIR sensor

1 Introduction

Forests mainly maintain the balance of the earth's ecology. Figure 1 shows the forest reserves in a million hectares among the top 10 countries. However, forest fires are significant concerns that damage this very own ecology [1]. The main problem with forest fires has been the inability to detect them early and may only be identified when the fires have caused significant damage. The delay also causes a problem in stopping the fire. It would have caused irreparable damage to both the environment and the ecosystem when it was stopped. On a long-term basis, this would have significantly affected our climatic conditions, the pattern of weather, global warming, and even the extinction of rare indigenous flora and fauna [2].

A. K. Singh (✉) · S. M. Rafeek · P. S. Harikrishnan · I. Wilson
School of Engineering, UPES, Dehradun, India
e-mail: singh.akshi85@gmail.com

N. A. Siddiqui et al. (eds.), *Advances in Waste Management*, Lecture Notes in Civil Engineering 301, https://doi.org/10.1007/978-981-19-7506-6_3

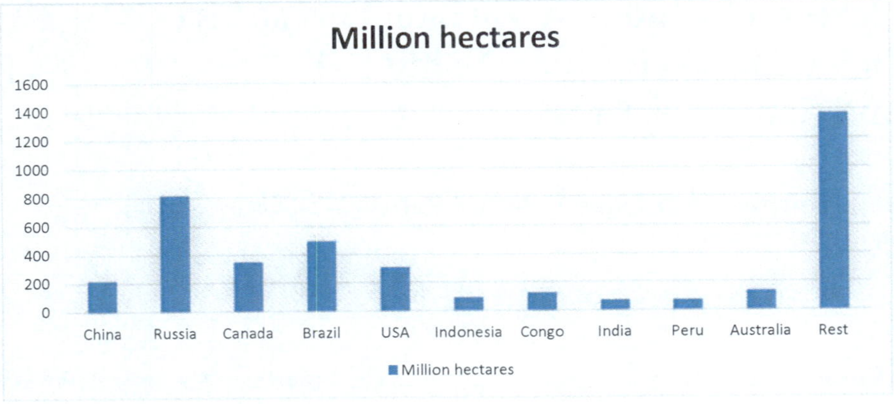

Fig. 1 Forest in a million hectares around the world [3]

Hefty fossil fuel usage and deforestation have decreased the forest's life, posing an existential threat to humans and other life forms. Indeed, our mother earth is diagnosed with a forest fire disease [4]. Researchers identifying this disease have initiated many afforestation programs; however, this would levy a heft charge upon them, estimated to be 5 trillion US dollars. Furthermore, discarded cigarettes and power arc lines are the major causes of this disease, whereas lightning is a natural cause. Thanks to significant technological advancement, we can identify lighting strikes early [5]. Many people have cited that this usage of technology involves a lot of money. However, it is essential to note the other side of the coin: the loss of lives and property damage. In this paper, we have reviewed three methods of detecting forest fires. It is important to note that increasing detection time would increase the time loss for stopping the fire. Figure 1 highlights the percentage of occurrence of fire in India. 54% of the fire happening in India owes high incidence, which means the county is more prone to severe fires. So indeed, it is necessary to identify the fire at its earliest to ensure its extinguishment at the earliest (Fig. 2).

1.1 *Bandipur Forest Fire Incident*

One of India's major forest fire outbreaks happened in 2019 at Bandipur National Park, located in Karnataka, India. The fire began on 21 February 2019, and continued its havoc until it reached a halt on 25 February 2019. The fire had damaged 4419.54 hectares of land during its reign. The fire was worsened by strong winds, making it very difficult for firefighters to control it. The Indian air force and the firefighting department played a significant role in stopping the fire.

It was reported that it had been two years since the fire was seen in Bandipur National Park. It was indeed disheartening to know that a countless number of reptiles and other small animals had lost their lives along with thousands of trees; however,

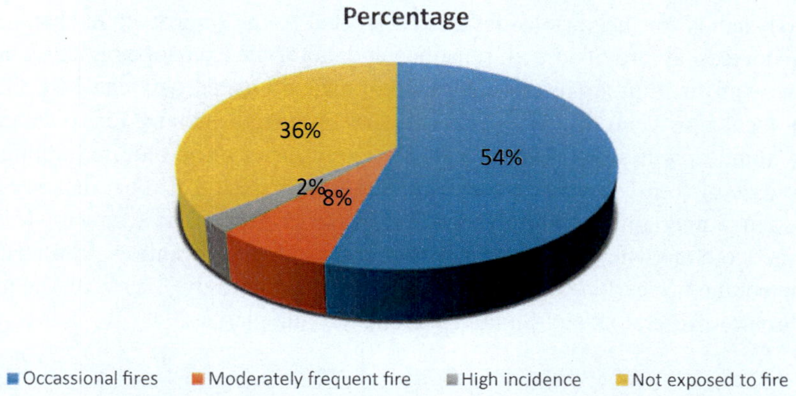

Fig. 2 Occurrence of fire in India [6]

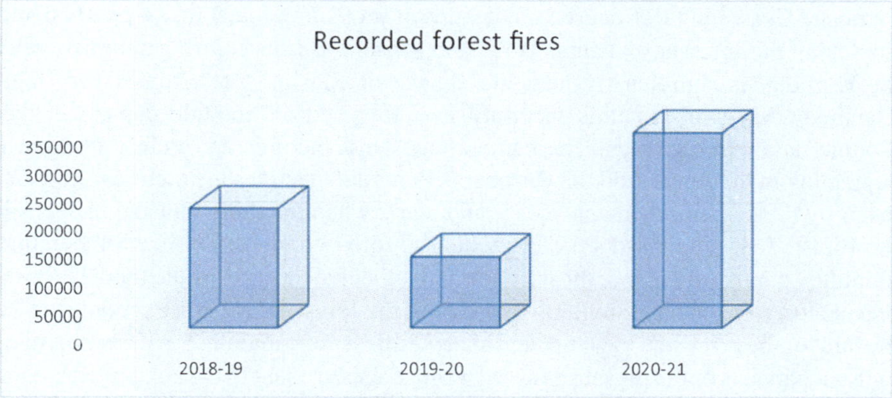

Fig. 3 Recorded forest fires in India [7]

larger mammals and fast-moving animals had found refuge elsewhere (Bandipur Blaze Was a Result of Bad Fire Management Policies, Say, Researchers, n.d.). Furthermore, on analysis, it was found that the sudden change in climatic conditions and the presence of a large amount of dry fauna could have been the reason for the fire to spread (Fig. 3).

1.2 Internet of Things (IoT)

The Internet of Things (IoT) is a new innovative concept used to observe, control, and intuitive automation. It involves any physical objects interconnected using a series of sensors, nodes, and displays for data processing and finally giving the desired results [8]. By 2020, about 30 billion devices were running over IoT's concept

[9]. IoT stands for the various devices connected to the Internet to exchange data. These devices are provided with sensors and data for the exchange of data. The data involves information about the environment. The collected data can be processed either locally or centrally using the network of systems. Many IoT is controlled using mobile phones or tablets, which remotely turn on the required application. In the case of forest fire detection, a small device is used to collect data regarding the rise in temperatures, humidity, and even pressure drops. These data are accessed anytime from anywhere and obtained in real time. IoT corporations in a forest fire can detect the fire early and take necessary steps accordingly [10]. With the help of IoT, Smart refrigerators and Smart TV came into the picture.

2 Literature Review

Patra and Goswami [10] focused on the idea of protecting forest fires by early detection using the Internet of Things (IoT) and sensors. Because during summer, when the humidity is significantly less, the chance of causing forest fire is very high. The methodology used in this paper utilises a Raspberry Pi module, camera, Zigbee module, and other sensors. The camera takes live pictures and videos of what is happening in the forest, and the Raspberry Pi module handles the pictures and video taken by the camera, then analyses it and alerts when anything unusual happens in the forest. Also, the Raspberry Pi module absorbs other factors like temperature, humidity, quality of air and, most importantly, the live location of the module for the precise locations in case anything happens in the forest. The Zigbee module acts as the hub for these sensors to gather and analyse all the information. So in case anything unusual happens or the preset values, alert messages are sent to the concerned authorities along with videos and photographs. Various research and development activities are going on for forest monitoring data analysis. In 2021, Patra and Goswami [10] utilised Internet of Things (IoT) to make the forest digitalised to improve the forest ecosystem. So the term forest digitalisation defines implementing new and latest technologies to improve existing methods. So Internet of Things (IoT) plays a massive role in improving these technologies. Incorporating IoT is to detect forest fires and illegal logging activities. The methodology is that all the sensors are attached to the trees in various areas. Then these sensors collect information like temperature, wind speed, humidity, and the intensity of light, and all the information collected is then sent to the Central Analysis Node, which analyses the information like the intensity of the fire, the direction of fire, and speed of the spreading of the fire. This information is then sent to a central hub located in the forest which is then stored in the cloud and can be accessed by the concerned authorities to take necessary control measures to prevent the spreading of the forest fire. Seamless connectivity, real-time monitoring or sensing equipment, and energy ingathering are the significant areas recommended in this paper to make the forest digitalised.

Sharma et al. [11] proposed a multi-model approach based on Internet of Things (IoT) and deep learning for identifying, propagating, and analysing live fire locations in agricultural fields and their respective activities. Here, the IoT used is the combination of sensors like various detectors and deep learning methodologies. Combining these various sensors or detectors is done with the help of fuzzy logic, thus helping give live time or precise location of the fire. The deep learning method uses IP cameras to provide reliable and precise data. The overall system also uses software to detect the live location of fire for determining the precise locations. Also, the software used has a wide variety of features like extracting the exact location of the fire, the name of the fire victims, and automatically sending the information to concerned authorities.

However, one of the most challenging areas in the present situation is the development of smart cities. In 2020, Sharma et al. [12] focused on implementing fire detection using Internet of Things (IoT) and image processing methods to sustain Smart cities. In contrast, for developing a Smart city surrounded by agricultural areas, forest areas, and more open vast areas where the probability of fire incidents and a threat to human life is very high, it is necessary to protect the surrounding environment from these fire threats. So IoT plays an essential role in detecting the forest using sensor networks and Unmanned Aerial Vehicle technology. The proposed framework consists of many wireless sensors, Unmanned Aerial Vehicles, and Cloud computing. Along with these sensors and devices, images are analysed and refined to detect the precise location of the fire.

In 2021, Patra and Goswami [10] proposed those essential things for improving and maintaining the agricultural domain using sensor networks and Internet of Things. So the paper mainly focused on what the primary equipment and components are, new and latest technologies that should be incorporated, issues regarding the security of the framework, challenges, and finally what the future scope of these sensor networks and IoT (Internet of Things) are in the field of agricultural activities. Nowadays, IoT and sensors are utilised in almost every field like Smart cities, forest fire detection, and healthcare services. With the increase in population, the demand for food also increases exponentially, so it is necessary to maintain the quality and quantity without any compromise. The concept of implementing IoT and wireless networks is emerging faster.

3 Different IoT Methods

3.1 Method 1

This method used an ESP32 board [13] that incorporates Wi-Fi and Bluetooth and has a more extensive range. It also used different sensors like rain sensors that detect rainfall and a sound sensor for detecting noise around the place. Another DHT 11 (temperature and humidity sensor) is used, which observes any rise in temperature and humidity around the forest. A passive infrared sensor (PIR) was also used for motion

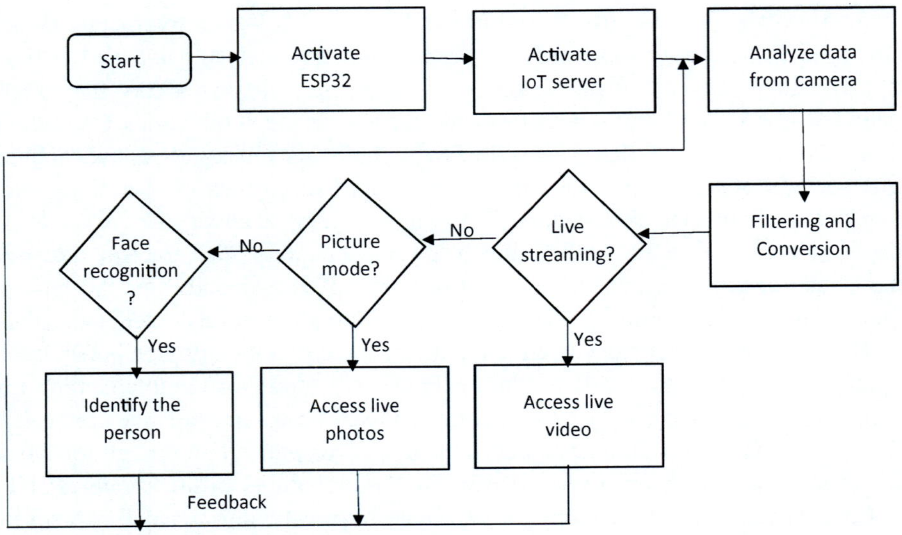

Fig. 4 ESP 32 camera flow chart

detection in the forest. Another ESP32 Camera module is used for face recognition and images of the surroundings. A step-down transformer was used as a power supply for the ESP32 board and buzzer. The buzzer gets activated when the threshold value is crossed, and an liquid crystal display (LCD) screen is available to get the data readings. The sensors collect the data and get verified from ESP 32 microcontroller [5]. The reading is then compared with the threshold reading and is activated when it exceeds the given value by the surrounding conditions of the atmosphere. It would then activate and alert the department to approach the surroundings for intimidation (Fig. 4).

3.2 Method 2

The system consists of interconnected and distributed amongst themselves and with other servers. Sensors involved are of different kinds that can sense humidity, temperature, and other required factors. IoT devices, user applications, and web interfaces are the main components that comprise the proposed system's entire architecture. Likewise, the device senses real-time data and is processed on time. The main aim is to inform the emergency services regarding the occurrence of the incident. However, the information is passed in two ways to improve efficiency: the sensor and the data transfer. The devices used sense data of the surrounding atmosphere and the magnitude of the meteorological constraints, and also gases in the air; the information is then transmitted using elements of the interface. The visualisation models are graphs, pictorial representation, and statistical reports.

Monitoring and controlling pollutant gases can help prevent the forest from fire accidents. It helps to determine the risk areas of the forest and prevent them by reducing the risk. For example, first, the temperature increased by 30 °C, and humidity decreased lesser than 30% in the forest area defined as a danger zone. This condition is informed through an alert message to the management [14]. The reason is that meteorological conditions are favourable to fire generation in the forest. Second, the pressure determines the period of storms/anticyclones that aggravate weather conditions, such as fire. Then, the excessive amount of CO_2 and CO concentration increase in temperature, and less humidity is the evidence of biomass fire [15]. This way, the sensor devices can determine the risk of fire accidents and generate an alarm to the nearest nodes and routers. It helps to identify and detect the fire in the forest and the location of the remote environment.

Fire accidents can be prevented by observing and regulating gaseous pollutants, which identifies the risky areas of the forest and help to prevent them. Moreover, when temperature increases by 30 degree Celsius and humidity also gets reduced by 30%, the area is called a danger zone. This condition would be informed to the management by alerting them of a chain of events. The message will be conveyed if the meteorological conditions are favourable for the occurrence of the fire. Secondly, pressure is another factor that would help aggravate the condition in case of a mishap like a fire. Incidents like cyclones and storms would help. An increase in the concentration of CO and CO_2 in the atmosphere would increase and decrease towards temperature and humidity, which eventually increases the risk of fire. Therefore, the sensors used would help guide the alarm to the nearest emergency station and help to act immediately following a fire, so that the forest regulatory authorities and other concerned authorities can take action immediately before the fire starts to spread at a faster rate (Fig. 5).

3.3 *Method 3*

The primary target of the paper is to identify and predict forest fire more swiftly to reduce the hefty loss that comes up due to it. This method comprises sensors with a wireless network that would help identify forest fires. The entire system proposed consisted of many closely connected sensors amongst themselves. These nodes collect the data like temperature and relative humidity and send this for processing using the cluster of nodes that collaborate amongst themselves and connect them using a neural network [16]. It is then used to produce the weather index, which predicts the climatic conditions to cause a fire. A sink is then utilised to transmit the message to the higher authority, which concludes using other factors to check whether there is any other probability of a fire occurring. The module could also give an emergency like smoke or sudden abnormal temperatures. In this scenario, the message is quickly sent to the central system. In the past, many neural detections using satellites were already in place [17]. In this proposed project, the author is the first to use in-network processing. The complex data network obtained is simplified

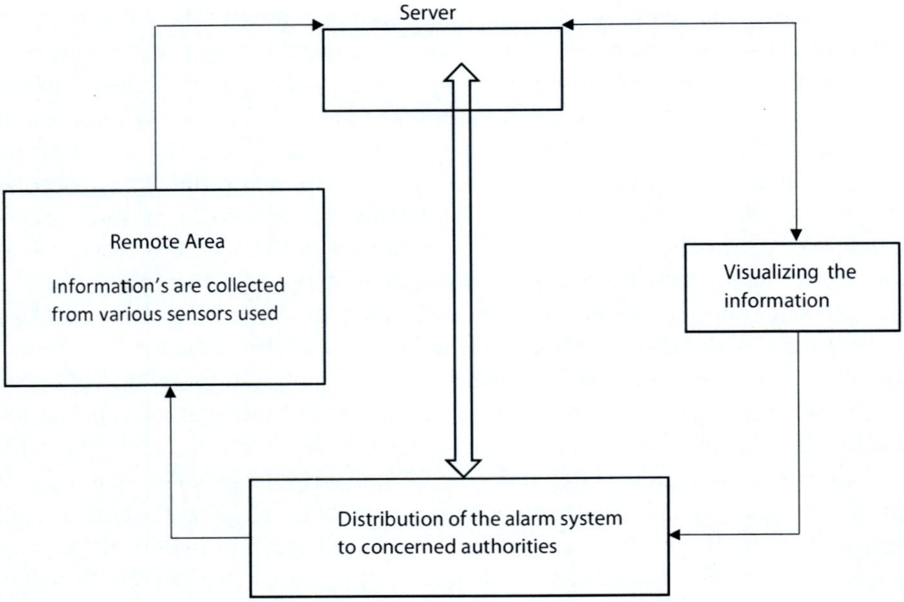

Fig. 5 IoT framework

by calculating the given data at every node. The main advantage is the ability to conserve energy and overhead communication by using only the tiny amount of data required for collectively obtaining valuable information for the final decision [18].

4 Conclusion

Nowadays, a lot of research and development is taking place to develop IoT and sensor networks further, because the application of IoT is limitless and currently used in a wide variety of fields like smart homes, smart cities, smart grids, agricultural fields, automobile sector, pharmaceutical field, and for the early detection of forest fires like what we have discussed in this paper. The framework they are using is almost similar in every methodology when it comes to IoT. However, the primary challenge we face while incorporating such a complex system of networks is the difficulty in quickly identifying an error. Also, it would become challenging to rectify when an error has happened due to its complexity. Another issue that arises as we use IoT is security and privacy. Furthermore, as technology advances, further incorporation of the latest innovations will enable us to get a more simplified system with better facilities and fewer chances for errors.

References

1. Halofsky JE, Peterson DL, Harvey BJ (2020) Changing wildfire, changing forests: the effects of climate change on fire regimes and vegetation in the Pacific Northwest, USA. Fire Ecol 16(1). https://doi.org/10.1186/s42408-019-0062-8
2. Alkhatib AAA (2014) A review on forest fire detection techniques. Int J Distrib Sens Netw 2014. https://doi.org/10.1155/2014/597368
3. Singh R, Gehlot A, Vaseem Akram S, Kumar Thakur A, Buddhi D, Kumar Das P (2021) Forest 4.0: Digitalization of forest using the Internet of Things (IoT). J King Saud Univ Comput Inf Sci. https://doi.org/10.1016/j.jksuci.2021.02.009
4. Deforestation causes. WWF
5. Kanakaraja P, Syam Sundar P, Vaishnavi N, Gopal Krishna Reddy S, Sai Manikanta G (2020) IoT enabled advanced forest fire detecting and monitoring on Ubidots platform. Mater Today Proc 46, 3907–3914. https://doi.org/10.1016/j.matpr.2021.02.343
6. Welcome To Forest Survey of India
7. India records highest number of forest-fire alerts in three years: Govt.
8. What is IoT (Internet of Things) and How Does it Work?
9. State of the IoT 2020: 12 billion IoT connections
10. Patra G, Goswami L (2021) Forest protection using wireless sensor network and IoT. Mater Today Proc https://doi.org/10.1016/j.matpr.2021.03.742
11. Sharma A, et al (2021) IoT and deep learning-inspired multi-model framework for monitoring active fire locations in agricultural activities. Comput Electr Eng 93:107216. https://doi.org/10.1016/j.compeleceng.2021.107216
12. Sharma A, Singh PK, Kumar Y (2020) An integrated fire detection system using IoT and image processing technique for smart cities. Sustain Cities Soc 61:102332. https://doi.org/10.1016/J.SCS.2020.102332
13. Development Boards (2022) Espressif Systems. https://www.espressif.com/en/products/devkits. Accessed 11 Feb 2022
14. Satendra and Kaushik (2014) Forest fire disaster management
15. De Sario M, Katsouyanni K, Michelozzi P (2013) Climate change, extreme weather events, air pollution and respiratory health in Europe. Eur Respir J 42(3):826–843. https://doi.org/10.1183/09031936.00074712
16. Akyildiz IF, Su W, Sankarasubramaniam Y, Cayirci E (2002) Wireless sensor networks: a survey. Comput Netw 38(4):393–422. https://doi.org/10.1016/S1389-1286(01)00302-4
17. Application of remote sensing and GIS for forest fire susceptibility mapping using likelihood ratio model—Geospatial World
18. Yu L, Wang N, Meng X (2005) Real-time forest fire detection with wireless sensor networks. Proc 2005 Int Conf Wirel Commun Netw Mob Comput WCNM 2:1214–1217. https://doi.org/10.1109/wcnm.2005.1544272

Strategies for Improving Municipal Solid Waste Management: A Comparative Assessment for Medium-Sized Cities

Ankur Chowdhary, Vibhor Goel, and Hitesh Pandit

Abstract Growing population and urbanization have exacerbated municipal solid waste management challenges globally. Most municipalities in medium-sized cities are experiencing performance challenges. Existing methods for collecting, transporting, and disposing of solid waste, in India as well, are in chaos. Rather than identifying and implementing proactive strategies to eliminate performance gaps in critical components, governments should focus on providing effective emergency responses. The objective of this study is to do a comparative analysis of performance indicators to evaluate the existing solid waste management system. This study looked at ways to strengthen the existing status of solid waste management in any medium-sized city in the direction of a cleanliness survey under the Clean India Mission. Effective responses to successfully implement source segregation of waste, door-to-door collection, decentralized processing, scientific treatment, benefiting informal/formal waste pickers, and street sweeping, based on particulars collected, need to be formulated to improve the environment, public health, and aesthetics. On average, only 21.6% of waste is segregated at the household level along with 14% of the operational capacity of compost plants in India, which is not appreciable; quick responses should be taken, which can be replicated for any medium-sized city.

Keywords Clean India Mission · Cleanliness survey · Municipal solid waste management · Performance assessment · Improvement strategies

A. Chowdhary (✉)
Foundation Cluster, School of Design, University of Petroleum and Energy Studies (UPES), Dehradun 248007, India
e-mail: achowdhary@ddn.upes.ac.in

V. Goel
C-2117, 12Th Avenue, Gaur City 2, Greater Noida West 201009, India

H. Pandit
Faculty of Planning, CEPT University, Ahmedabad 380009, India

N. A. Siddiqui et al. (eds.), *Advances in Waste Management*, Lecture Notes in Civil Engineering 301, https://doi.org/10.1007/978-981-19-7506-6_4

1 Introduction

In 2018, cities around the world created 2.01 billion tonnes of municipal solid waste (MSW), averaging 0.74 kg per person per day [1] (Table 1). World municipal solid waste generation is anticipated to increase by 70% to 3.4 billion metric tonnes by 2050 [2]. Various causes, such as population expansion, increased earnings, and changing consumer patterns, all contribute to this variation in solid waste generation Planning [3]. Developed and developing countries have very different waste management practices; there is a huge gap between their practices. Since the 1960s, various experts have been accommodating developed countries' Solid Waste Management (SWM) agencies [4].

In India, solid waste management (SWM) is a major concern for many urban local governments. It has become a major challenge for the authorities to manage. India has a population of 1.31 billion inhabitants, with 31% residing in urban areas [5]. By 2030, India will have 68 cities with more than a million population and 590 million people will live in urban areas [6]. Urban India is grappling with the ever-increasing task of meeting the rising population's infrastructural needs. The growth in the generation of solid waste is one obvious consequence of the increasing population. In 2014, 143,449 tons per day (TPD) municipal solid waste (MSW) was generated in India [7], which increased to 152,076.70TPD in 2019 [8], out of which 149,748.6TPD (98%) of the total municipal solid waste was collected, while only 55,759.6TPD (37%) were processed or treated.

The large-scale concerns about inadmissible SWM lead to the release of the MSW (Management & Handling [M&H]) Rules, 2000, which was further revised in 2016 requiring all urban local bodies to establish a proper waste management system [7]. The Ministry of Urban Development launched the Swachh Bharat (Clean India) Mission (SBM) in 2014 which intends to address the challenges in managing the MSW and to help urban areas in creating present-day and suitable frameworks [9]. To evaluate the performance of urban local bodies (ULBs) in cleanliness, sanitation, and hygiene, an annual survey called Swachh Survekshan (cleanliness survey) under SBM Mission was rolled out on the parameter of service-level progress structured in

Table 1 Regional waste generation (annual)

S. No	Region	Waste generation (Million Tonnes)
1	East Asia and The Pacific	468
2	Europe and Central Asia	392
3	South Asia	334
4	Latin America and The Caribbean	231
5	North America	289
6	Middle East and North Africa	129
7	Sub-Saharan Africa	174

Source What a waste 2.0: a global snapshot of solid waste management of 2050 [1]

the Swachh Survekshan toolkit. Despite the guidelines and encouraging models, most ULBs continue to face difficulties with on-site segregation, collection, transportation, treatment, and scientific disposal of waste, resulting in environmental and public health damage [10]. For the design of an efficient SWM system, regular monitoring and data collecting are required, to improve SWM practices throughout the country by creating a centralized database based on ULB's SWM experiences and employing modern tools and technologies such as remote sensing, GIS, and mathematical optimization [2]. Municipal Solid Waste Management (MSWM) study must take into account socio-economic, environmental, financial, and institutional factors, as integrated approaches are a viable tool for addressing the current waste management issue in developing nations.

The Wasteaware ISWM (integrated sustainable waste management) benchmarking indicator is a method for assessing the effectiveness of a city, municipality, or group of municipalities' municipal solid waste management and recycling systems in a standardized manner. This method is derived [11] for the complete performance monitoring of SWM's physical components as well as governance issues, taking into account all important stakeholders.

This paper focuses on strategies based on the performance assessment of solid waste management in one of the urban centers in Uttarakhand State, Rishikesh, having a population of 106,320 [5]. Rishikesh Nagar Nigam has 40 wards spreading over an area of 11.56 Sq Km. According to Rishikesh Municipal Corporation, the city generates approximately 58.84 MT of solid waste every day. The indicators for the assessment are as per the cleanliness survey toolkit. The Assessment of the current situation provides a Ground for the formulation of strategies, which can improve upon the situation of Solid waste management in Rishikesh. For Service Level Progress, two parameters are considered, i.e., *Collection and Transportation* (nine indicators) and *Processing and Disposal* (13 indicators), each indicator for both the parameters is associated with some marks which decide the progress of the ULBs [12].

Based on these two categories, individual scoring on various indicators has been done through which the overall score on service-level progress of Rishikesh city is calculated and by providing the strategies on which the cleanliness can achieve and service-level progress score can increase.

2 Materials and Methods

The application of performance indicators (PI) may have different purposes depending on the type of entity (regulation, managing entity) that uses them. The process of this study is divided into four parts listed below (Fig. 1).

First, the scope of performance assessment is defined in the context of solid waste management, in which various. As the current study focuses on the performance assessment of the SWM of Rishikesh city through the process suggested by Cleanliness Survey [4, 13–15] based on which performance indicators will be reviewed and summarized. In the second step, reliable data is collected to visualize the relevant

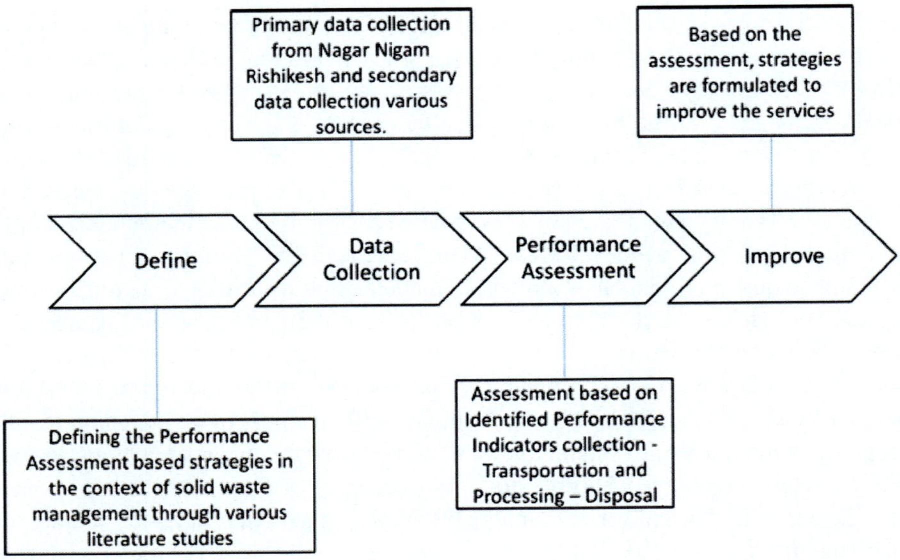

Fig. 1 Process flow chart

set of identified performance indicators, major parameters for the PIs will be collection and Transportation, and Processing and Disposal, based on which performance assessment can be done. In the end, strategies are formulated to improve the services.

3 Result and Discussion

3.1 Performance Evaluation of ULB Through Service-Level Progress

The Cleanliness Survey has become the most appropriate measure for performance monitoring, database management, and benchmarking of urban local bodies. The number of ULBs assessed has progressively increased over the years. Minister of Housing and Urban Affairs conducted a cleanliness survey to evaluate the performance of 4237 cities in January 2019 [12]. For service-level progress of solid waste management of Rishikesh city, the following are the parameters mentioned, under which it is being evaluated.

3.2 Collection and Transport

The nine indicators of collection and transport based on which performance evaluation is done are listed in Table 2.

Source Segregation of waste

Source segregation becomes the first and the most important stage in the entire solid waste management cycle. Currently, source segregation is not appreciable; on average, the source segregation is 5.8% only within the study area. There are no separate bins in the household for waste segregation into biodegradable and non-biodegradable, as per the SWM Rules, 2016. Street debris collected through pushcarts are also non-containerized which leads to the mixing up of waste.

Door-to-door Collection

The door-to-door collection plays a major role in collecting the domestic and commercial waste in the study area; currently, 16 auto tippers are collecting the waste in the municipal area. There is only 56.84% coverage by the municipalities. Currently, there is no agency, which is being impaneled for the collection and processing of waste in the study area.

There are 32 secondary collection points with 39 dustbins with a total capacity of 72 MT (14 dustbins of 3.5 cum capacity and 25 dustbins of 4.5 cum capacity). As per the buffer analysis Ref. [16] at the secondary collection, points to calculate the serving areas shows that 71.19% of the city is covered.

ICT-based Monitoring Mechanism

As there are only 16 tippers, which carry waste in the city, the tippers have no digital monitoring system, which creates a transparency issue with the current system and affects efficiency. Garbage-vulnerable points are also not being monitored which creates a nuisance to the public health and the environment, and affects the aesthetics of the city.

Informal Waste Pickers and Benefits Extended to Formal and Informal Waste Pickers

None of the informal waste pickers are formally integrated into a sustainable livelihood by Nagar Nigam. During the study, it was found that none of the sanitary workers including sweepers were wearing personal protective equipment which directly influences their health. Besides, there was no initiative for capacity building of sanitary workers and informal waste pickers.

Street Sweeping and Public Cleaning

Currently, road Sweepers are carrying out sweeping manually at around 68 km of major city roads and 223 km of local roads. For every 1 km of Minor road, one sweeper should be deployed Ref. [16]. In the study area, there are 130 street sweepers, therefore there is a gap of 93 sweepers without any mechanical equipment. The

Table 2 Performance evaluation based on collection and transport

S. no	Collection and transportation		Scheme of ranking					Current status	Total score
1.1	Door-to-door collection	Percentage of wards covered with operational Door-to-Door Collection of waste	>50%	50–64%	65–79%	80–95%	>95%	40	100
			20	40	60	80	100		
1.2	Source segregation of waste	Percentage of Wards practicing source segregation of waste which is maintained till processing/disposal site	>50%	50–64%	65–79%	80–95%	>95%	25	125
			25	50	75	100	125		
1.3	ICT-based monitoring mechanism	ICT-based monitoring mechanism in place for 1 Ward-wise collection and transportation (C&T), Collection from gates, monitoring of garbage vulnerable points (GVPs), and sanitation staff	ICT-based monitoring of collection and transportation of waste from all Gates (Door)	Are all gates (doors) monitored through ICT-based system	ICT-based GVP monitoring	(GVP should be monitored for at least 15 days from the date of transformation)		0	40
			10	10	10	10			
1.4	Informal waste Pickers	Percentage of Informal Waste Pickers formally integrated into Sustainable Livelihoods	No. of waste pickers formally integrated with ULB, SHG, NGO, etc	60–79%	80–95%	>95%		10	35
			10	20	30	35			

(continued)

Table 2 (continued)

S. no	Collection and transportation		Scheme of ranking					Current status	Total score
1.5	Benefits extended to formal and informal waste pickers	Benefits extended to all Sanitary workers including Informal Waste Pickers, i.e., workforce engaged under/through Jagirdari system, SHG, NGO, private agency, and informal waste pickers	PPE to all workers	Linkages established with at least two eligible Government Schemes	Monthly recognition of best-performing workers	Training imparted to more than 90% of workers		0	30
			7	7	8	8			
1.6	Street Sweeping and Public Cleaning	Public cleaning: 100% Wards are Clean in the Urban Local Body (ULB)	Twice a day sweeping all commercial areas	Once a day sweeping all residential areas	Transformation of all GVPs	All wards are bin-free	Up-keeping of slums and old city	0	50
			10	10	10	10	10		
1.7	Stormwater drains	Are Storm Water Drains and Water Bodies in all wards clean?	>50%	50–74%	75–95%	>95%		0	40
			10	20	30	40		0	40

(continued)

Table 2 (continued)

S. no	Collection and transportation		Scheme of ranking						Current status	Total score
1.8	Plastic waste management	Plastic Waste Management Rules: Has the City banned single-use plastic including plastic with < 50 microns during all festivals/social gatherings/events?	No action is taken	Only ban notified			Yes, ban notified, enforced, and fines collected		30	30
			0	15			30			
1.9	3R principles	3R Principles: Have measures been taken to reduce the generation of Dry/Wet Waste? If yes, share details	0	1	2	3	4	5	0	50
			0	10	20	30	40	50		
		Total (A)							105	500

frequency of street sweeping in commercial areas and residential areas of Rishikesh city is once a day and on an alternate day basis, respectively. Illegal Dumping is also Prominent in the city, which leads to garbage vulnerable points.

Stormwater Drains

Half of the ward peripheries in the city is having uncovered drains mostly leading to the Ganga River. Most drains in the municipal limits are dumped with dry waste such as plastic which leads to clogging.

Based on these indicators, the Collection and Transport parameters of the Cleanliness survey toolkit have been scored in the table below.

With the calculated score, of 105 out of 500 in the collection and transportation performance indicator, it is exhibited that the performance of the municipal area is not satisfactory and needs up-gradation to become a clean and green city.

3.3 Processing and Disposal

The second parameter for service-level progress has 13 indicators (refer to Table 3) which are explained as follows.

Landfill Site

The Integrated Solid Waste Management concept is closely linked to the 3R approach (reduce, reuse, and recycle), which is also aimed at optimizing MSW management from all the waste-generating sectors (households, commercial, and institutional establishments) and involving all the stakeholders (waste generators, service providers, informal sector, regulators, government, and community or neighborhoods) [7]. Currently, there is no scientific treatment and disposal of waste in Rishikesh city. The 3R principles are also not applied in the city, without any in-house composting by residents. This implies that the city is not within the lines of an integrated solid waste management Hierarchy. One of the prominent things that have happened in the city is that the city administration has banned the use of plastic bags and collected fines from the defaulters.

All the collected waste goes to Govind Nagar Dumpsite having an area of approximately 10 Acres. On the site, rag pickers are used as the labor required for Bio-mining or bioremediation. Currently, wet waste and dry waste processing capacities are insufficient to cater to the current waste. Apart from this, there is no strategy to treat Construction and demolition waste.

After conducting the Buffer analysis [16] for the location of the landfill site, it was found that the dumpsite is unfit for disposing of the waste. At present, the height of the squander is 3–4 m considering the height limitation of 15 m; the life span of the landfill site is around 17 years. It is found that the current location of the landfill site is in the buffer zone of Ganga River and National Highway, i.e., 120 and 200 m, respectively, which needs to be strategized.

Table 3 Performance evaluation based on processing and disposal

S. no	Processing and disposal	Scheme of ranking						Current status	Total score
2.1	Is the capacity of wet waste processing facilities in the city matching with the total wet waste generated by the city?	<50%	51–60%	61–70%	71–80%	81–100%	100%	0	50
		0	10	20	30	40	50		
2.2	Percentage of wet waste generated is actually processed, either by decentralized or centralized facilities	<40%		40–59%	60–79%	80–95%	>95%	0	150
		0		75	100	125	150		
2.3	Percentage of generated dry waste (excluding plastic and domestic hazardous waste) collected is actually processed/reused/recycled, either by decentralized or centralized facilities	<40%		40–59%	60–79%	80–95%	>95%	10	60
		10		30	40	50	60		
2.4	Percentage of total plastic waste collected is treated/reused/recycled, either by decentralized or centralized processing	<40%		40–59%	60–79%	80–95%	>95%	5	40
		5		10	20	30	40		
2.5	Percentage of total domestic hazardous waste collected is treated, either by decentralized or centralized processing	<40%		40–59%	60–79%	80–95%	>95%	0	30
		10		15	20	25	30		

(continued)

Table 3 (continued)

S. no	Processing and disposal	Scheme of ranking						Current status	Total score
2.6	Is there any mechanism in place to manage Construction and Demolition (C&D) waste as per C&D Waste Management Rule, 2016? Are plans in place to initiate the processing of C&D Waste?	Land identified and Plan in place for processing C&D waste		Dedicated area(s) earmarked to Keep C&D waste in the city	User charges for services and fines being collected for open dumping	Dedicated vehicles in place	C&D Waste Helpline in place	0	50
		10		10	10	10	10		
2.7	Remediation of existing dumpsites undertaken and the stage of the same or no legacy waste (dumpsite)	No process started	The tender called/published	Agreement signed	60–79% of waste remediated	80–95% of waste remediated	>95% of the waste remediated or no dumpsite/legacy waste	5	60
		0	5	30	40	50	60		
2.8	Is the landfill in the city a sanitary landfill? Or landfill not required/Zero landfill city	No process started		Tenders called for the construction of a sanitary landfill site	Agreement for construction done but work not commence	Sanitary landfill under construction	Sanitary landfill available and being used/Landfill not required	0	50

(continued)

Table 3 (continued)

S. no	Processing and disposal	Scheme of ranking					Current status	Total score
		0	5	15	40	50		
2.9	Percentage of Bulk Waste Generators (BWG), including those generating more than 100 Kgs (or less as notified by the State/city) of waste per day, practicing on-site processing of their wet waste or outsourced to private agencies—processing not outsourced to ULB	Less than 40%	40–59%	60–79%	80–95%	>95%	10	50
		10	20	30	40	50		
2.10	Has city empanelled service provider(s) managing collection and processing of dry/wet waste to cater to Bulk Waste Generators (BWGs) or households not being covered under Door-to-Door Collection?	No			Yes		0	30
		0			30			
2.11	Percentage of households processing their wet waste at Home/Community Level (Households under RWAs will qualify under the BWG definition)	<1%	1–2%	2–3%	3–4%	>5%	0	50
		0	20	30	40	50		

(continued)

Table 3 (continued)

S. no	Processing and disposal	Scheme of ranking					Current status	Total score
2.12	Percentage of Swachhata App/Local App complaints cover issues related to littering/garbage dumping/overflowing litter bins	>25%	16–25%	11–15%	6–10%	< 5%	0	40
		0	10	20	30	40		
2.13	What percentage of the operational cost of Sanitation and Solid Waste Management is covered by Property Tax (SWM/sanitation subhead), User Charges (for SWM/sanitation-related services), Sale of city compost, and Advertisement rights on CT/PT and Litter Bins?	<40%	40–59%	60–79%	80–95%	100%	0	40
		0	10	20	30	40		
	Total (B)						30	700

Bulk waste collection and Transport

In the study area, there are five bulk waste generators, generating 8.4 MTPD waste that is segregated into 56% biodegradable and 44% non-biodegradable waste. Bulk waste generators are required to treat their wastes inside the premises but are not able to do that because they do not have that kind of onsite composting facility.

Grievance redress

The Municipal Council of Rishikesh uses IVRS for receiving complaints regarding Municipal issues. The complaints received from citizens are from various domains such as drain cleaning, littering, and garbage dumping.

Finance

As per the Annual Operation and Maintenance cost of Rishikesh Nagar Nigam, it is in debt of 5 crores. Total expenditure in the solid waste management of Rishikesh is Rs.646.2 Lakhs/year with revenue of Rs.147.36 Lakhs/year through user charges and challans only. We can see that only 23% of the SWM expenditure is being covered through the revenue and the rest relies on central and state grants.

By analyzing expenditure and revenue collection, it can be seen that coverage of user charges is about 32.4%. Other revenue sources like recycling materials, composting, RDF, etc. are null.

As per the current status described above, the processing and disposal indicator of the Swachh Sarvekshan toolkit has been scored in the table below.

Through the performance evaluation matrixes steer as per the guidelines of Swachh Sarvekshan toolkit 2020. If we add the score of both the indicators, i.e., Collection and Transport, and Processing and Disposal, Rishikesh scores 135 (A + B) marks out of 1200.

As we can see, the total score of Rishikesh Nagar Nigam is not satisfactory; it requires some Performance Improvement Strategies, which can also be generalized for the One Lakh Plus populated cities, to have a cleaner city, better performance index, and Swachh Sarvekshan ranking.

4 Performance Improvement Strategy

For collection and transport, the wet, dry, chemically inert, and domestic hazardous waste should transport through auto tippers from households, commercials areas, institutions, and public places to the Decentralized Material Recovery Facility (MRF) out of which wet waste should go for composting at Composting centers and dry waste shall be segregated Manually by workers at Decentralized Material Recovery Centers. The compost produced through wet waste can be sold to fertilizer companies and dry waste to recyclers/intermediates, which will also benefit the revenue part.

Chemically inert and domestic hazardous wastes left at Decentralized MRF are then segregated and sent to sanitary landfill and Domestic Hazardous Waste Landfill sites separately.

4.1 Collection and Transport

Door-To-Door Collection of Waste

The major issue lies in the low coverage of Door-to-Door collection due to the Jagirdari system and people's careless attitude, not giving waste on time, and improper dumping of waste at secondary Collection Points. Therefore, the concept of bin-free city was devised.

Phase 1: Contribute your waste to make your city bin-free

Awareness about the bin-free city will be created among the people and with the removal of big bins, the people will be urged to throw waste in the auto tippers only. (However, twin bins will be there every 250 m; 282 bins are required up to 2021, 336 up to 2031, and 356 up to 2041. Auto tippers will empty these bins in day time whereas Refuse Compactors Deployed for BWG at Night.)

Phase 2: Involvement of Informal Collectors.

Informal collectors working under the Jagirdari system are responsible for the waste collection from the household. Due to their low coverage and unsound way of waste collection, they will be involved in the Formal way of waste collection as a driver, helper, or they might be linked to skill India or NULM through ULB.

Phase 3: Give your waste and win Prizes.

In this phase, a value will be created in waste so that people would find a benefit in getting attached to the formal waste stream; there would be three Stacks of Lottery Prizes, namely 30 days, 27–29, and 25–27 days of giving waste under the formal system. This way, the public will be encouraged to give waste under the formal system.

Phase 4: Linking User Charges with the property Taxes.

The last step would be linking the property taxes with the user charges such that people will pay for the services even if they do not give waste, which in turn will make people liable to take the service.

Apart from this four-phase strategy, NGOs and SHGs could also be involved in the collection process to improve efficiency.

Segregation at source

The significant issues are addressed through the Segregation strategy devised as a proposal for the Plan. This strategy will have four phases:

Phase 1: Information, education, and awareness.

In this phase, through wall paintings and jingles, awareness of waste segregation will be raised. After that, the mayor or active political leader will communicate individual messages to every household about waste segregation, followed by the distribution of bins for proper containment.

Phase 2: Incentivization or Segregation Crediting.

In this phase, if a waste generator segregates waste for a day, he will get a coupon worth Re.1. The coupon can either be used to rebate the user charges at the same rate or can be used to get a rebate over property taxes at the same price, and lastly, he could also use that coupon for Purchasing the Goods. The Commercial shop owner can again use this for any of the things; it will be a reusable coupon circulated just like regular currency.

Phase 3: Penalization.

The waste has to be segregated into three bags provided by Nagar Nigam. If a person is not segregating the waste, he will be fined 10 Rs. per day and his fines can be waived off if he segregates waste for the next ten days.

This strategy will not be applicable to bulk waste generators as the wet waste will not be taken.

Secondary Collection and Transportation

Secondary Collection and Transportation will be done temporarily in the Study Area, the timeline for which is subjective to whenever the Segregation and Door-to-Door Collection reach the range of 90–100% or up till 2021 (whichever is less).

Since the Secondary Collection and Transportation did not cover some newer parts of the city, therefore, some of the Bins from the Location of Excessive capacity were shifted to the location not having them.

ICT-based Monitoring Mechanism for ward-wise Collection and Transportation

All the gates will be installed with RFID tags, which will be scanned by the helper during the door-to-door collection. The helper will check whether the waste is segregated or not and accordingly record it in the app, simultaneously the waste generator will also receive a message This way the waste generators can also keep a tab on the helpers; it will be two-way monitoring.

The vehicle used for waste collection can also be monitored using GPS.

Apart from this, geotagging and manual monitoring of all the Garbage Vulnerable points must be there. Their transformation must also be looked upon, and continuous monitoring must be their post-transformation.

Informal and formal Waste Pickers and Benefits extended

All informal waste collectors should be identified and brought under the formal system of solid waste management. Informal waste pickers and rag pickers should be asked for their other potential livelihood options, thus linking them with Government policies to create a livelihood for them.

Apart from this, PPE should be distributed to all the informal sanitary workers (informal/formal), and creating a competitive spirit by introducing a Best Sanitary workers award could also be a way to improve those workers' efficiency.

Annual Training Programs must be launched for training all the sanitary workers, including all the sweepers and sanitation staff.

Street Sweeping and Public cleaning

Rishikesh road sweeping is not done at Beat Length, which is a major issue as there is Ineffective Sweeping and also an Issue of Alternate Sweeping due to a shortage in the workforce that can be observed in some areas.

It was assessed that rather than providing Sweepers on National Highway and City Main Roads, a mechanized Sweeping would be much more beneficial, whereas sweepers will be deployed at a beat length of 1 km on the Local Roads.

Sweeping in commercial areas must be increased to twice a day, whereas for residential areas, it must be at least once a day. Apart from this, all garbage vulnerable points must be identified and monitored for at least 15 days post transformation.

The community bins should be removed since it is not being used appropriately; the bins have littered waste around them. The removal of community bins and the 90% coverage of door-to-door collection will help make the city bin-free.

Collection and Transportation of waste in slums must also be done to maintain the sweeping service, and collection from these areas must be improved.

Whether Storm Water Drains and Water Bodies in all wards are clean

All the drains should be covered with perforated grills to prevent clogging, and they should be cleaned at least once a week.

3R Principles

Implementation of Extended Producer Responsibility:

- Conversion of flower waste to Incense Sticks.
- Issuance of Crockery Bank from Nagar Nigam free of cost for social and public events.
- Toiletry kept in the hotel should be fixed/dispensed or available with a minimum charge.
- Artifacts made out of waste material, second-hand sale of old books/hardware/other material, and Waste Exchange Program, 'Neki Ki Diwar' to leave goods/articles for needy ones.
- Donating the waste food to gaushalas and needy people.

4.2 *Processing and Disposal*

A decentralized approach will be used for processing the waste [16]. The option of waste to energy will not be adopted because this approach could be put to use only when the waste inflow will be greater than 500 TPD.

Wet waste Processing—Decentralized Composting

A decentralized way of Composting will be much more beneficial. The Feasibility of different types of composting is shown in the table below.

Technology	Windrow composting	Vermicomposting	Aerated static pile	In-vessel
Min waste required (Sqm.)	100	0.1	0.1	0.5
Min area (Sqm.)	12,500	100	300	200
Capex (Lakhs)	650	0.25	2.5	2.5
Opex (PA) (Lakhs)	70	1.8	1.8	2.16
Compost generation (TPD)	15	0.04	0.04	0.15
Operating cost (Rs/Tonne)	750	5000	1800	1700
Emissions	Yes	No	No	No
Leachate	Yes	No	Yes	No
Landfill	Yes	No	Yes	Yes

As the decentralized composting will be done, it should be kept in consideration that no emissions are produced while Composting; hence Windrow and aerated static pile composting will not be adopted [17]. If we consider the remaining two, Vermi-composting and in-vessel Composting, options, the latter will be a feasible option in financial terms. There will be five Composting Zones in the study area, and all the wet waste produced in these zones will be composted within these zones.

The capacity of the composting facility at dedicated sites in five zones is equal to the waste generation, i.e. 28.68 MTPD. The promotion of In-house composting should be done in an incentivized manner by using the user charges' incentives as a tool.

Dry waste Processing Decentralized MRF

The city would have five Material Recovery Facility (MRF) centers and drum Composting Centers. Here, the segregation of Recyclables into paper, plastic, metal, clothes, Rags, etc. will be done manually. These MRF centers are proposed based on population. *The percentage of Plastic Waste is dominant over other categories. Therefore, all road developers must utilize the same amount of plastic in the bitumen mixes to construct roads within the 50 km periphery of the city* [18].

Plastic will be Shredded and bailed then given to the road contractors, and paper will be bailed and sold to paper mills; apart from this, thermoplastic, metal, clothes, and rags will be sold directly to Recyclers.

Balers and shredders are provided in adequate quantities to deal with the 25.77 TPD of dry waste generated.

Construction and Demolition waste

There is a need to develop a system for the management of Construction and Demolition (C&D) waste in the city. Stationary Crushing plants are only feasible for waste generation above 20 TPD, either on a cluster or individual basis. Since the waste generated here is below 20 TPD, mobile crushing units are being proposed. Waste after getting crushed will be stored at C&D waste storage.

The process of construction and demolition of waste begins with the segregation and storage of waste by the generator. The collected waste is then screened and crushed into secondary crushing, which after the screening gives coarse and fine aggregates, followed by the removal of dust and washing of sand, which gives sand, silt, and clay. The end product after processing the C&D waste is recycled aggregates and recycled concrete aggregates.

A helpline number will be issued which can be used by the Waste generator to call the authority for crushing the waste and taking it to the waste storage site; apart from this, User charges for mobile processing of C&D waste will be linked with the Building Permit and fines should also be issued to those who are dumping the C&D waste openly.

Disposal of Waste

After the segregation of waste at decentralized MRF into inerts and recyclables, the Inert and Domestic hazardous waste goes to the sanitary landfill site, and there will be a separate portion of the landfill for the Hazardous domestic waste.

The sanitary Landfill site will be on a cluster basis. The old Dumpsite is carrying a capacity of 1.21 Lakh MT, which needs to be Bioremediated or Bio-capped, as the odor of the waste in the locality is creating a nuisance; hence, it needs to be treated and land should be reclaimed and converted into a green area.

Bioremediation and Biomining:

In bioremediation and biomining, the whole land is reclaimed and the process takes 2–3 years. After the Immediate treatment, operation, and maintenance are required, whereas no money has to be spent on leachate treatment and gas extraction [19]. The municipal corporation can register 15 Ragpickers as a workforce in bioremediating and booming the old dump yard. Bioremediation and biomining will be done using the trench method [20].

Bulk Waste Generators

Wet waste generated by the Bulk waste generators should be treated inside their premises [16]. However, there will be an option from the corporation's side for Mobile composting on a Chargeable Basis. The Dry Waste will be taken once in two days from the bulk waste generators and collected through the Refuse compactors with a Frequency of two.

Grievance Redressal

As the community bins are being removed from the city, the littering issue will be reduced. Apart from that, smart solutions like monitoring through CCTV over the garbage vulnerable points and dumping sites or effective patrolling over these points reduce the dumping of waste.

Revenue Generation from waste recycling and composting

The recycling of waste and composting will help generate revenue. The following ways can be adopted for revenue generation:

- The compost could be sold at 3 Rs./kg (The [21]).
- The plastic waste could either be sold to road contractors or sent to the cement kilns for co-processing.
- Polystyrene, metals, and glass are present in low quantity hence they can be sold through Intermediates.
- The old clothes could be sold either to the Intermediates or the textile recyclers.
- User charges, challans, and BWG processing fees will also form a major component in recovering the cost of operations and maintenance.

5 Conclusion

Rishikesh city is witnessed as a tourist attractive place with a large-scale urban expansion. The Current load on SWM infrastructure is failing the city. There is no proper monitoring of the solid waste management system, which leads to the unaccountability of sweepers and informal workers. The current mechanism for the collection and transportation of waste is lacking in coverage and monitoring without any innovation and public awareness. The processing and disposal of waste are lacking in the proper method of treatment. Based on the Swachh Sarvekshan toolkit, various indicators are covered, provided that Rishikesh city can only score 135 marks out of 1200 which indicates the pathetic condition of SWM in the city.

To increase the score for better ranking in Survekshan, strategies which are provided can increase the scorecard to a much higher number. The expected score can change from 135 to 1075 marks, which is quite high. These strategies are based on the levels prescribed in the indicators of the Swachh Sarvekshan toolkit. Following these strategies, Rishikesh municipal corporation and other 1 lakh + ULBs could improve upon their Swachh Sarvekshan ranking as well as the service delivery in the solid waste management domain to have a healthy and clean city.

References

1. Kaza S, Yao L, Bhada-Tata P, Van Woerden F (2018) What a waste 2.0 a global snapshot of solid waste management to 2050. World Bank Group. https://doi.org/10.1596/978-1-4648-1329-0
2. Singh S (2019) Solid waste management in urban India: imperatives for improvement. J Contemp Issues Bus Gov 25. https://doi.org/10.47750/cibg.2019.25.01.001
3. Planning Commission (2014) Report of the task force on waste to energy: Volume I prepared by Planning Commission (PC). Task Force Waste to Energy I:1–178
4. Byamba B, Ishikawa M (2017) Municipal solid waste management in Ulaanbaatar, Mongolia: systems analysis. Sustain 9:6–9. https://doi.org/10.3390/su9060896
5. Census (2011) Census [WWW Document]. www.censusindia.gov.in
6. Sankhe S, Vittal I, Dobbs R, Mohan A, Gulati A, Ablett J, Gupta S, Kim A, Paul S, Sanghvi A, Sethy GM (2010) India's s urban awakening : Building inclusive cities, sustaining economic growth. McKinsey Q 1–33
7. CPHEEO-Part I (2016) Swachh bharat mission-municipal solid waste management manual part I: an overview. CPHEEO (Central Public Heal. Environ. Eng. Organ. Minist. URBAN Dev., India, pp 1–62
8. CPBC (2019) Solid Waste Management Rules 2018–2019. Annual Report 24
9. CPHEEO (2016) CPHEEO (Central Public Health and Environmental Engineering Organisation), Swachh Bharat mission-municipal solid waste management manual II: The Compendium. Ministry of Urban Developemnt (201
10. Joshi R, Ahmed S (2016) Status and challenges of municipal solid waste management in India: a review. Cogent Environ Sci 2:1–18. https://doi.org/10.1080/23311843.2016.1139434
11. Wilson DC, Rodic L, Cowing MJ, Velis CA, Whiteman AD, Scheinberg A, Vilches R, Masterson D, Stretz J, Oelz B (2015) "Wastewater" benchmark indicators for integrated sustainable waste management in cities. Waste Manag 35:329–342. https://doi.org/10.1016/j.wasman.2014.10.006
12. MoHUA (2020) Swachh Survekshan-2020
13. Armijo C, Puma A, Ojeda S (2011) A set of indicators for waste management programs. In: 2nd International conference on environmental engineering and applications, vol. 17, pp. 144–148
14. MoHUA (2018) Swachh Survekshan Survey Toolkit 2019. Toolkit
15. Sequeiros B (2012) Application of performance indicators in municipal waste services. Fenix Tecnico Ulisboa Pt 1–10.
16. SWM-Rules-2016.pdf, n.d
17. Thakkar Sunny A, Yadav Bhupendra C, Patel Ronak K, Shaikh Matin R (2016) Solid waste management by composting–a feasibility study. 479–482
18. MoRTH (2015) Use of Plastic waste
19. Datta S, Eastes L (2015) Biodegradation in Municipal Solid Waste landfills [WWW Document]. https://www.geoengineer.org/education/web-class-projects/cee-549-geoenvironmental-engineering-fall-2015/assignments/biodegradation-in-municipal-solid-waste-landfills
20. Abhishek Kumar A, Li J, Khan J, Pandey A (2019) An overview of the potential of bioremediation for contaminated soil from municipal solid waste site. Emerg Eco-Friendly Approaches Waste Manag
21. The Hindu (2015) Compost from city waste ready for sale. The Hindu

Microbial Interactions with Accessories—A Complete Analysis

B. C. Arundhathi, S. Prasanth, R. Sivaranjani, P. Abishek, A. Arumugam, and Sudalai Subramani

Abstract "Connecting People", "Make Believe", "Rule the change", "Practicality as the priority, and win with technology", "Nothing Like Anything", "A Better Life, A Better World", "Explore beyond limits", "Think different", "Yours is here", "Be more", "Move On", "Beyond Precision", "The Pen the world prefers", "The Joy of Writing", and "The power toothbrush more dentists use", these are some of the famous taglines or slogans we hear in our routine life. This shows the outreach of the accessories in the life of human beings in their daily routine. With the advent of these gadgets and accessories, the interactions with the human hand are increasing and therefore the microbes present in the skin can transfer to these gadgets and accessories and then to another human who may handle them. The common microbes found in the accessories and gadgets are bacterial species like *CoNS, Diphtheroids, Micrococcus species, Bacillus species, Propionibacteria, Alpha streptococci, Viridans streptococci, Pseudomonas species,* and fungal species like *Aspergillus niger, Aspergillus flavus*, Yeasts, and Molds. The micro-organisms are resistant to antibiotics like Ampicillin, Cefoxitin, Erythromycin, Clindamycin, Cefuroxime, Ciprofloxacin, Cotrimoxazole, Linezolid, and Vancomycin by Antibiotic Susceptibility Testing. Therefore, the study gains its importance that the micro-organisms gain susceptibility to the antibiotics so that the research studies have to be extended for the treatment of diseases caused by the microbes isolated from the contaminated gadgets and accessories as the user are increased year by year.

Keywords Microbial contamination · Electronic gadgets · Accessories · Nosocomial infection · Disease caused

B. C. Arundhathi · S. Subramani (✉)
Centre for Pollution Control and Environmental Engineering, School of Engineering and Technology, Pondicherry University, Kalapet, Puducherry 605014, India
e-mail: sudalaijothi@gmail.com

S. Prasanth · R. Sivaranjani · P. Abishek · A. Arumugam
Bioprocess Intensification Laboratory, Centre for Bioenergy, School of Chemical and Biotechnology, SASTRA Deemed University, Thirumalaisamudram, Thanjavur, Tamil Nadu, India

N. A. Siddiqui et al. (eds.), *Advances in Waste Management*, Lecture Notes in Civil Engineering 301, https://doi.org/10.1007/978-981-19-7506-6_5

1 Introduction

The keenness of observing the microbial world for the human being starts in the seventeenth century when Antonie Van Leeuwenhoek observed micro-organisms using a microscope of his design. Initially, microbiology was applied to the medical field and later it expanded to various fields like Pharmaceutical, Industrial, Agricultural, Veterinary, Environmental, Water, and Aerobic Microbiology. The history of clothing dates back 170,000 years ago and personal hygiene care products with their low-tech predecessor form were available in the distant past. But the form which is used nowadays was available in the twentieth century using the available materials which give attention to contemporary life. According to the Czech philosopher Radovan Richta, the emergence of technology can be classified into three stages viz., the first stage is the creation of a tool that provides the mechanical advantage in accomplishing any work, and the second stage is the creation of machines which is a tool that lessens the human effort and it needs only the control of its function, and the third stage is the automation which is a machine that avoids human effort by its automatic algorithmic programming language. Due to the advent of machines, gadgets, and accessories or personal care products in the twentieth century, the need to link with microbial studies is essential as they are one of the sources of microbial contamination to humans [8]. This review paper tells about microbial contamination through gadgets like Mobile Phones, Computer Keyboards, Earphones, and Accessories like Contact Lens, Pens, Watches, Gloves, Catheter, Stickers used for Intravenous lines, Insulin syringes, ATMs, Toothbrush, Jewel-like Rings, Shoes, Stethoscopes, Forceps, White Coat, Medical Charts, Necktie, and in places where there is more hand contact like Door Handles and Knobs, Faucet Handles, Stretcher Handles, and Wheelchairs. Some of the journals taken for this review paper include "The American Journal of Infection Control", "Journal of Parenteral and Enteral Nutrition", "British Journal of Infection Control", "Journal of Infection Prevention", "Journal of Association of Physicians of India", "Journal of Infection in Developing Countries", "Journal of the American Veterinary Medical Association", "Indian Journal of Dental Research", "International Journal of Preventive Medicine", "Open Infectious Diseases Journal", "Medical Journal of Australia", etc. The search engines used are "Scopus", "Science Direct", "PubMed", "Google Scholar", "PLOS One", etc., and the keywords used are "Microbial contamination", "Personal hygiene", "Electronic gadgets", "Personal care products", "Disease caused", "Nosocomial Infection", "Death rate due to hospital-acquired infection", etc.

2 The Factors Responsible for Contamination in Accessories

Microbes are present virtually everywhere on Earth. The growth of microbes depends on the physical parameters (osmotic pressure, temperature, and pH) and chemical elements (carbon, oxygen, and nitrogen sources). Because of their adaptability and size, microbes exist in any environment and occupy only a small space. The requirement of nutrients is less and diverse for microbial growth. Therefore, microbes can also be present in electronic gadgets and accessories which provide the optimum conditions to grow.

3 Materials and Methods for Isolation of Micro-organisms

The collection of samples is the foremost step in the process of finding micro-organisms from them. The most common method used for the collection of samples from accessories and gadgets is using a sterile swab. For the collection of samples, the sterile cotton swab was moistened with normal Saline or Sterile water or Trypticase Soy Agar [28], and then it is rolled on all the sides of the surface of the sample collected. The saline is most readily available and easy to wet the dry cotton swab rather than others, therefore it is frequently used. Then the cotton swab is streaked on the plates of different media according to the prescription like Blood agar, Mac Conkey agar [70] (Ta et al. n.d.), or Blood of sheep mixed with Levine's formulation [Virender Singh et al.,] or Nutrient agar [Mbajiuka Chinedu et al.,] or Sabouraud Dextrose Agar [Yazhini Jagadeesan et al.,] or Brain Heart Infusion agar [Sue Elizabeth Shajan et al.,], or Thioglycolate medium, Neomycin blood agar with Metronidazole 5 μg disk [63]. The swab was made of Nylon and to moisten it, buffer (Tween 20, 0.15 NaCl) was used. Using this swab, the sampling was done on the surfaces [41]. Growth of the samples was monitored during incubtion at 37 °C up to 48th hour. The incubation provides the ideal environment for the micro-organisms to grow. Morphological identification of the organisms, color, and characteristics of the colony, Gram staining, Sporulation, Motility, biochemical tests includes Catalase test, Oxidase test, and Sugar fermentation. In addition to this, dextrose utilization and nitrate reduction, and Voges-Proskauer test are also done (Ta et al. n.d.). From all the above tests, the pure culture of the microbial species whether bacteria or fungi or protozoa with their species was identified. Antibiotic susceptibility testing (AST) is carried out to determine the resistance of bacterial species for a long time. The testing for antibiotic sensitivity is often done by the Kirby-Bauer method. The antibiotic sensitivity test includes methicillin resistance and multidrug-resistant species [74]. The common antibiotics used for antibiotic sensitivity tests are Ampicillin, Cefoxitin, Erythromycin, Clindamycin, Cefuroxime, Ciprofloxacin, Cotrimoxazole, Linezolid, and Vancomycin for *Staphylococcus aureus* species [Asheeka Zainab Arif et al.].

4 The Accessories

The things which are inevitable in our life are Mobile phones, Computers, Pens, Wrist Watches, Contact lenses, Rings, Toothbrush, Helmets, Public utilities like Door Handles/Knobs, Wheelchair handles, Medical charts, Stethoscopes, and attires like Doctor's White Coat, Necktie, etc., which also shows microbial contamination.

4.1 Electronic Accessories

4.1.1 Keypad of Mobile Phones

One of the devices which is a boon and curse at its time is the Mobile Phone. Despite having many advantages over the phone, the disadvantages are also accumulated like a heap. Rapid and increased cell phone usage makes the individual more sedentary, with increased leisure behavior, mental stress, abortion in case of pregnant ladies, and increased risk of Glioma and acoustic neuroma. Apart from these disadvantages, the hidden and virulent curse is the presence of micro-organisms in mobile phones. Mobile phones are used to bridge the gap between the individual who needs the medical facility with the adjacent healthcare facility. Thus, the mobile phone became an essential accessory for healthcare workers. But the healthcare workers are not properly cleaning their hands and mobile phones routinely which will lead to nosocomial infection [63]. From the hands of the healthcare workers, the microbes got transferred into the accessory which may then cross-contaminate with others while handling. The micro-organisms isolated from the mobile phone and hands of the healthcare workers such as doctors, nurses, and healthcare staff are *S. aureus* (52 and 37.7%), and gram-negative strains (31.3 and 39.5%), respectively. *S. aureus* is methicillin-resistant whereas gram-negative strains are Ceftazidime-resistant. The nosocomial isolates are Staphylococcus (33.3%), Non-fermentative gram-negative strains (21.4%), Coliforms (21.4%), Enterococci (7.1%), and Yeasts (11.9%). The microbes which are isolated from the hands and mobile phones were similar which reveals the microbes can transfer from one hand to the mobile phone and then to another case [70]. The micro-organisms isolated are skin flora and are resistant to commonly used antimicrobials such as amoxicillin (25% resistance), Cephalothin (6.7%), and Gentamicin (15.6%) which shows the clinical importance of spreading the nosocomial infection [59]. The efficacy of the alcohol-based hand rub in killing the microbes was checked and it worked well to some extent in which after disinfection with this rub, the hand has no/little microbes whereas the count increases after touching the mobile and fixed phone to 33/40 and 38/40, respectively [76]. From different studies, a wide range of micro-organisms such as anaerobes, coagulase-negative staphylococci, *Pseudomonas aeruginosa, Bacillus subtilis, Staphylococcus epidermis,* meticillin-resistant *S. aureus, Klebsiella snippets.* are present on the surface of mobile communication devices [8, 34, 53, 58, 72] (Ta et al. n.d.). The antimicrobials checked for the resistance of micro-organisms were Oxacillin [58], Cephalothin, Amoxicillin, Gentamicin, etc. [59].

4.1.2 Computer Keyboard and Mice

"The sole function of education is to open the way to thinking and knowing, and the school, as the outstanding organ for the people's education, must serve that end exclusively"—Albert Einstein. Computers play a major role in educational purposes and are frequently in contact with human hands for every action to be performed. Computers are used in every place such as hospitals, schools, colleges, and shops. The computer keyboards in the dental setting can be subjected to contamination by the routine handling for the access of patients' medical and dental records. The common micro-organisms found in the computer keyboard of a dental setting are *S. aureus* and Staphylococcus. The life span of *Candida albicans* on desktop keyboards is 10 days [50]. The analysis of computer keyboards was done with four components: (1) Contamination degree, (2) Impact of ethanol on cleaning, (3) Transmission of bacteria to the keyboard from gloves, and (4) Individual's hygiene [22]. The computer keyboards act as a reservoir of pathogenic micro-organisms, and they lead to various outbreaks of diseases and also lead to nosocomial infections from the medical staff to patients [19, 75]. The common micro-organisms isolated from the computer keyboard and mouse are bacilli, Pseudomonas, and staphylococci strains (Fraser and Girling, 2009). The domain where the isolation of micro-organisms taken is Dental setting [50], consulting rooms in veterinary services [21], Veterans Affairs Medical Center [57], University hospital [14], healthcare setting [55], Intensive care unit (ICU) setting [9], Burns Hospital [47], etc. The disinfectant called Isopropanol proves to be efficient in that there is a 96% reduction of viable micro-organisms [50]. Despite the alcohol-based hand rub proving efficient in reducing micro-organisms, users are reluctant to use the hand sanitizer often. Therefore, the suggestive measures can be voice-recognition software, washable keyboards, and virtual keyboard images [64]. Incorporation of a polymer with antimicrobial activity with polyurethane reduces the strains of methicillin-resistant Staphylococcus aureus (MRSA) (97.8%), vancomycin-resistant Enterococcus faecalis (VREF) (95.0%), Escherichia coli (99.99%), and Pseudomonas aeruginosa (92.1%) at a concentration of 0.5% at 240 min [16]. 2% Chlorhexidine Gluconate with 70% isopropyl alcohol (CHG) also proves to be efficient in the decontamination of microbes [29]. Disinfection is more efficient in the case of ICU when compared to medical wards where more patients are handled, and this insists the rate of spreading of micro-organisms increases when the contact is more [15].

4.2 Communication Devices

4.2.1 Earphone

There is an exponential increase in mobile phone users since 2017, and the users of mobile phones are forecasted at 4.77 billion ("Mobile phone users worldwide 2013–2019|Statistic," n.d.). The mobile phone is coming up with so many facilities in which one of the facilities is listening to music with the use of the earphone. Students and college students frequently use earphones which can act as a reservoir of micro-organisms that will cause Otitis externa. The microbial analysis of the ear and earphone samples for the users with infrequent use and those who use it at a stretch of 30 min and frequently in sharing also shows that the earphone samples of frequent users are around a 68% growth when compared to infrequent users who are around 56%. This result shows the importance of maintaining personal hygiene practices even when sharing the earphone with others ("OJHAS: 2008-2-4, Mukhopadhyay C et al. A comparative analysis of bacterial growth with earphone use," n.d.). A biofilm is a group of micro-organisms that can be present on any surface that causes various disease outbreaks [35]. The sharing of earphones is not only a personal hygiene problem but also a problem concerning health status of the people ("Isolation and characterization of biofilm formation of microbes from Children's nails, ear, earphones, and feeding bottles|Iram Liaqat—Academia.edu," n.d.). Thus, much concern should be made when sharing the earphones.

4.2.2 Hospital Bed Handsets

The Hospital Bed Handsets are prone to microbial contamination in which 103 out of 115 samples were contaminated with *coagulase-negative staphylococcus, Bacillus species, fungal species, non-hemolytic streptococcus species, a-hemolytic Streptococcus species, Staphylococcus aureus, and methicillin-resistant Staphylococcus aureus* [78].

4.3 Handles

4.3.1 Door Handles/Knobs

The Door Knobs present in public premises were subjected to the high prevalence of micro-organisms because of the frequent contact with many hands-on surfaces. The different microbial flora present in different circumstances can lead to cross-contamination with other individuals who are constantly handling these surfaces. In the Abuja metropolis, these surfaces were investigated for microbial contamination.

Out of 180 swab samples collected and cultured, 156 (86.7%) were positive. Contamination rate was higher in commercial areas than the Government offices, and banks where most of the bacterial contaminants were Coliforms [48]. A suggestive measure can be the use of hand sanitizer in public places where more people are gathering. The door containing powder-coated AgION silver zeolite was confirmed using scanning electron microscopy that it considerably reduces the microbial growth over the steel surfaces against all the tested conditions, and therefore it can be used as a remedial measure ("The door containing powder-coated Agion silver zeolite—Google Search," n.d.).

4.3.2 Hand Rims and Handles of Wheelchairs

Wheelchairs used for the transportation of patients who are unable to walk are things where there is more hand contact. Therefore, the Wheelchair handles in hospitals are contaminated [25]. Contamination of *S. aureus* on wheelchair hand rims was higher than on handles but ethanol wipe disinfects the whole strain. Therefore, regular wiping with ethanol is recommended to avoid cross-contamination among healthcare workers.

4.3.3 Ambulance Accessories

The Ambulance care assistants are frequently handling the stretchers for moving patients who require medical care during emergencies, or accidents with the help of Ambulances. As the usage of the ambulance is often, the study gains importance. The sample was taken from oxygen flow meters and stretchers before and after fumigation. Growth of culture was evident on the swab and was vulnerable to fumigation. Therefore, the fumigation technique offers better decontamination against microorganisms whereas the chemicals used in this technique are toxic to most forms of life, including humans. Therefore, suggestive measures which don't harm humans should be developed [5].

4.3.4 Handles of Shopping Trolleys

Increasing Shopping Malls are changing the lifestyle of people where the products can be purchased by the individual with the help of shopping trolleys. But the problem accounting for this is the multiple handling of the trolleys by which the transfer of microbes from one surface to another happens. Isolation of MRSA and *S. epidermidis* from Handles of Shopping Trolleys in which 11 isolates of MRSA and 70 isolates of *S. epidermidis* were obtained from a total of 340 handles collected and cultured proving the confirmation of microbes in the trolley handles.

4.4 *Medical Accessories*

4.4.1 Stethoscope

The Stethoscope is used to test the internal sounds of the lung and heart of an animal or human body, and it is an important accessory carried by healthcare professionals. The stethoscopes obtained from healthcare workers at MGM Medical College, Navi Mumbai, before and after the use of alcohol rub at the diaphragm of the stethoscope show that 90 (90%) of stethoscopes showed bacterial contamination which is statistically significantly different and the isolates were *Staphylococcus aureus* (56%) followed by *Bacillus species* (42%), *Micrococci* (24%), *CoNS* (4%), *Pseudomonas aeruginosa* (4%), *Diptheroids* (4%), *Enterobacter species* (2%), and *Candida species* (2%). Thus, doctors should follow hygiene practices to avoid the transfer of microbes from one to another and also protect themselves, Gurjeet Singh et al. Many strains are well built to resist the antibiotics increasing the infection. Poor disinfection procedures among doctors and other healthcare workers cause 98% contamination that will lead to nosocomial infection and also cause morbidity and mortality [4, 62, 71].

4.4.2 Medical Charts

An interesting study was conducted on the taint Medical Charts which is the collection of medical records of various patients in a hospital. It was regularly handled by the nurses, patients, and HCW. The study was conducted at various wards of the hospital located in Taiwan. The medical chart is not only a collection of patients' details about their medical conditions but also a reservoir of many pathogenic micro-organisms [37, 67]. The ID badges of the healthcare workers contain common respiratory or gastrointestinal viruses which are also nosocomial sources of infection [68].

4.4.3 Operating Room Boots

Operation theater is a facility in a hospital where surgical operations are carried out. The accessories used during the surgery and the attire worn by the persons inside the operation theater are thoroughly disinfected before entering the room. But improper disinfection of the boots may cause various diseases due to microbial transmission. Majority of the boots were contaminated with blood and bacteria. They are associated with skin microbiota or the environment. Contamination of the boots produced by Wellington was less than usual [2].

4.4.4 Catheters

Catheters are medical device that finds their application in many fields such as cardiovascular, urological, gastrointestinal, neurovascular, and ophthalmic applications. *Staphylococci* were the commonly found skin commensals on the surface of the catheter which will lead to Peritonitis caused due to continuous peritoneal dialysis [51]. Long-term intravenous catheters (LTIVC) are also subjected to microbial contamination [27, 38].

4.4.5 Attire Healthcare Workers

There is a significant opportunity for a patient to face microbial contamination by the white coat of healthcare workers [12]. Here too, *Staphylococci* are the commonly found microbes in the white coat of healthcare workers [69]. Contamination was higher in the pockets and cuff areas of the coats, and it majorly depends on the individual's usage of the white coat but not on the length of time. More contamination was seen in the coats of surgical personnel compared to medical personnel [76]. The white coats in the dental operatory had coats of 100% bacterial contamination [40]. White coats can also pave way for the transmission of pathogenic micro-organisms from one patient to another.

4.4.6 Elevator Buttons of the Urban Teaching Hospital Setting

The prevalence of colonization of elevator buttons was 61%, and there is no significant difference in colonization prevalence regarding the buttons' location or position of the panel. Staphylococci were cultured largely than the other micro-organisms which weren't frequently grown. Colonization on the buttons of an elevator is most common than on the surface of the toilets [31].

4.5 Accessories Used in Day-to-Day Life

4.5.1 Motorcycle Helmet

Helmets are becoming a compulsory accessory while traveling to avoid severe head injuries because the cost incurred on them is huge. The patients who wear helmets at the time of accidents sustain severe head injuries (4.8%) compared to persons who don't wear them (23.7%). 2.6% suffered critical injuries when worn with a proper chinstrap, whereas 14% of them are non-strapped ones [45]. The total expenditure incurred on head injury patients in a special ward, ICU, and during operation is around Rs. 1062/bed/day, Rs. 3082/bed/day, and Rs. 11,948/operated patient, respectively [42]. Thus, the wearing of a helmet gains importance. As the wearing of helmets gains importance, the point of linking it with microbial contamination also has a valid view in which the helmets are used for a long duration in which the sweat produced can lead to fungal contamination. The microbial contamination of commercial Motorcycle Helmet in the commercial city of Lagos, Nigeria, in which 300 motorcycle helmets were randomly collected and the species found in the helmets are species of Pseudomonas, Escherichia and Bacillus [1]. If hygiene is not properly followed, microbial transmission can cause health problems from the microbes transmitted from helmets. The helmets should be regularly cleaned with the disinfectant solution, and sharing the helmet should be avoided. Regular airflow should be there during the travel which avoids sweating thereby reducing the chance of growth of fungal contaminants inside the helmets.

4.5.2 Finger Rings

Rings are the traditional jewel worn by everyone, and it is the common one worn by healthcare workers including nurses during their patient handling and other related work. The effect of jewelry on the bacterial counts, the lowest bacterial counts being found on the control areas of skin which is adjacent to jewels worn, is increasing on the jewelry surfaces, and the highest counts on the skin are tested under the jewelry. Gram-positive bacteria such as *S. aureus* and *Enterococcus spp.*, gram-negative bacteria included Acinetobacter species, Enterobacter species, Pseudomonas species, Klebsiella species, *Stenotrophomonas maltophilia*, E. coli, *Serratia marcescens*, and *Proteus mirabilis* were found in the rings of the nurses in the ICU wards in which there is no significant difference between the plain wedding rings and the rings with stones. Also, hand sanitization with alcohol rub proves ineffective in the reduction of microbial load in hands [77].

4.5.3 Toothbrush Head

For maintaining good oral hygiene, everybody should brush their teeth twice a day as recommended by the Dental Association. The microbes can stay in the toothbrush for one week because of the humid conditions [3]. The germs can then enter the circulatory system through the oral cavity and then can cause even heart problems, Pneumonia, etc. [13]. The presence of hard bristle tufts describes the microbial contaminants such as the presence of *S.mutans, S. aureus, and bacillus species* in the toothbrushes present in the toilets which do not have bathrooms attached and depict that the micro-organisms can be sustained for a long time on the surface of toothbrush which leads to diseases like tooth decay and other ailments that badly affect the oral health of an individual [32]. Silver and its compounds are being widely used as an antimicrobial agent everywhere like face creams, silver-coated catheters, silver-incorporated bioactive glass systems, and silver-nanoparticle embedded antimicrobial paints [6, 36, 54]. The engineered method which is the silver coating of the toothbrush head to reduce the microbial intact doesn't decrease the microbial contamination also the presence of microbes such as *S. sanguinis* and *C. albicans* were significant when compared to the controls [3]. Cetylpyridinium chloride (CPC) proves the effective destruction of micro-organisms in air-dried brushes. The growth of *Staphylococcus epidermis* reduces to 100% and *Candida albicans* had a reduction of 94%, and it proves to be an economical and feasible solution [43]. The spray which is primarily used for the disinfection of toothbrush bristles that contains propylene glycol and distilled water [vehicles for formulation for other substances], methylparaben and propylparaben [preservatives], polyvinylpyrrolidone K30 [dispersing agent], ethyl alcohol [solvent], and chlorhexidine digluconate [active antimicrobial component] proves to be another effective solution against other sprays and sterile tap water for the reduction in the microbial growth such as aerobes, anaerobes, streptococci, and gram-negative bacilli [56].

4.5.4 Contact Lenses

Contact lenses are considered medical devices that can be worn to correct vision, for cosmetic or therapeutic reasons, and is prevalently used by college students. Unfortunately, the individuals are unaware of the contamination spread by the Contact Lens, Lens Case, and Lens Care Solutions due to poor hygiene practices. The most frequent user of contact lenses, i.e., among university students 41.79% of the samples collected were contaminated with microbes [39]. The *Staphylococci species* are commonly found in lenses which are worn. The contamination of contact lenses, lens solution, and lens cases may lead to keratitis which is a condition in which the eye's cornea becomes inflamed. Eighty-two (81%) out of 101 samples were contaminated with microbes where 77% of contact lens cases grew bacteria including *Staphylococcus epidermidis* and *Staphylococcus aureus*. Around twenty-four varieties of fungal contaminants are available and among that *Cladosporium species* were majorly found [65]. The development of biofilm was over the contact lens surface even after the disinfection. All the microbial contaminants like bacteria, fungi, and protozoa release an enzyme called catalase which will break down hydrogen peroxide into hydrogen and water. Continuous exposure to hydrogen peroxide may have selected antibiotic resistance [24].

4.5.5 Writing Pens

As the hands of healthcare workers are a major cause of cross-contamination, the materials handled by them are also to be undergone microbial analysis. One among them is the writing pens. The survival time is the duration of microbe survival in the sample. On rubber grip pens, *S.aureus* and CONS survived up to 72 and 48 h, whereas on plastic pens both species survived up to 48 h, and on metallic pens, they survived up to 24 and 18 h, respectively. The copper-containing pens have fewer micro-organisms compared to the stainless steel pens [11]. Pulmonologists teach, distinguish, and serve patients suffering from cardiac and respiratory problems, and therefore they had more chance of getting infections. This view was supported by the study in which bacteria were found in 17 of the 20 pens in which the species found were *Coagulase-negative staphylococci* in all the pens. The effective disinfection method for writing pens is to rub them with alcohol frequently which will greatly reduce the gram-positive cocci present in the hands of healthcare professionals and hospitalized patients [26]. The different disinfection procedures followed by the candidates to clean the pen are alcohol cotton, antiseptic, and wet wipes. The microbial contamination was more in female participants (72.5%) compared to male participants, and there is no bacterial contamination in the pens from the participants collected from the gynecology department rather than the Pediatric department. The time required to kill the micro-organisms is an important criterion for the disinfectant or the antimicrobial to act. The pens with *Micrococci spp.* don't show microbial growth at the 24 h, and it shows only after 72 h [66].

4.5.6 Necktie

The necktie can also have the incidence of acting as a vector for microbes, and the median bacterial load found in the necktie is ten times greater than the wound and the microbes isolated are *Staphylococcus aureus* (methicillin-resistant), *S. aureus methicillin-sensitive* (MSSA), *Enterococcus spp* and aerobic gram-negative bacilli [52].

4.5.7 Automated Teller Machines

The microbial interaction with the Automated Teller Machines with its antibiotic resistance pattern concludes that 95.7% of swabs were contaminated against 92 swabs collected from different locations such as door handles, ATM monitors, keyboards, and card swiping machines, and money outlets in ATM centers in and around Pondicherry [46].

4.5.8 Purses

The woman doctor's purses act as a vector of contamination by the fomites if she touches the colonized object and then the purse if hand hygiene is not performed. Around 70% of the purses were contaminated and therefore the accessory should be properly disinfected before handling by the patients [20].

4.5.9 Footware

Footware is for the protection of an important part of the body, the foot. The regular cleaning of the footware is uncommon, and there they harbor numerous microbes.

Species such as yeasts, Entero-coccus species, E.coli, and total bacteria were present in the athletes' shoes and due to improvements in sanitation processes the bacterial presence in athletes' shoes significantly reduced.

4.5.10 Vacuum Cleaner

The dust collected from the vacuum cleaner is responsible for the gastro-intestinal infection which is caused by Enterobacteriaceae and salmonella species. The following factors are used for the analysis of microbial species—the presence of pets, presence of shoes indoors, the model, time period of the vacuum cleaner, and dirtbag [61].

4.5.11 Data Analysis

After the data was collected, it was analyzed using the SPSS statistical software (Statistical Package for the Social Sciences) with various versions like 11.5, 13, 15, 17, 18, 16, 20, and 21, and also by using some statistical methods like Wilcoxon U-test, and χ^2 analysis using Fisher's exact test to find out the statistical significance of the presence of the micro-organisms isolated in the gadgets and accessories. If the presence of micro-organisms is significantly present in the gadgets and accessories, then it may lead to hospital-acquired or nosocomial infection in case of hospital environment.

5 Conclusions and Discussions

5.1 Hospital-Acquired Infection

Hospital-acquired infection otherwise called "Healthcare-associated infection" or "Nosocomial infection" is an infection acquired by patients after being admitted to the hospital. This kind of infection was acquired during the hospital stay and appeared to show its symptoms after discharge. The hospital-acquired infection increases only the stay in the hospital which brings no good to health [17]. It may be because of the poor hygiene practices of the healthcare workers and the hospital environment. The reason for the rapid increase of hospital-acquired infections are the crowded environment, frequent patient transfer between the wards, and the susceptibility of acquiring infection who has less immune power like newborn patients, burn patients, etc., according to Prevention of Hospital-Acquired Infections "A Practical Guide" 2nd edition released by World Health Organization [62]. The median total cost difference between Nosocomial infected and Non-nosocomial infected patients tend to be higher in the pediatric intensive care unit in Andalusian, Spain, during the year 2008–09 [44]. The micro-organisms commonly involved in hospital-acquired infection are *Staphylococci.*

5.2 Suggested Remedial Measures

The simplest remedial measure against the microbes is the maintenance of Hand hygiene. WHO emphasizes the importance of hand hygiene among healthcare workers who are constantly in touch with the contaminated surface. During healthcare operations, thousands of people around the world died due to insufficient hand hygiene practices. Thus, WHO initiated the "SAVE LIVES: CLEAN YOUR HANDS" campaign to insist on the significance of hand hygiene. The alcohol hand rub can be used after touching the contaminated sites as it proves more effective rather than other measures, for 20–30 s. If the hands are visibly soiled, then use soap and water to wash the hands for 40–60 s.

Despite mentioning the importance of hand hygiene during healthcare operations, compliance among healthcare workers is low and it is because of the low surveillance methods. Thus, the electronic audible alerts could be used as a suggestive measure in performing hand hygiene compliance among the workers during their shifts [73]. The commercial hand washes and soaps advertised as "antimicrobial" contain Triclosan. The minimum inhibitory concentration (MIC) defined as the lowest concentration at which no visible growth was observed, and was high for "Lifebuoy", "CAREX", and "Dettol" which are active against various organisms [30].

Copper can be incorporated into the surface of various equipment and fittings such as door handles, toilet seats, bed seats, and sinks, which proves efficient in reducing the microbial population such as surface micro-organisms, *E. Coli,* etc. [10, 23, 33]. The ionized form of Silver ($Ag + 1$) also proves to be efficient in antimicrobial resistance since silver is used for the purpose for ages [7]. The combined use of β-lactams and Daptomycin proves considerable reduction of Staphylococcus aureus, the most significant nosocomial pathogen [18].

The cleaning of the accessories like mobile phones, computer keyboards, and monitor screens is low, and therefore antibacterial screen protectors are invented to act against *E.coli Staphylococcus aureus.* The screen protectors prove to be simple and inexpensive means as an antimicrobial surface and the principal antimicrobial agent present in the protectors was crystal violet [49]. Despite various efforts to prevent microbial contamination, the environment of the hospital turns out to be a hotspot containing different pathogens.

The microbes present in the accessories are of great concern because of their increased handling for a long duration of time in our day-to-day lives. The skin micro flora can adhere to the surface of the accessories which will contaminate not only the accessories but also the individual's hands without any proper cleaning or disinfection of the accessories. The health impacts due to the micro-organisms are considerable because the strains are microbial resistant. To tackle antibiotic resistance, the 68th World Health Assembly in May 2015 endorsed the Global action plan on antimicrobial resistance. The Global action plan on antimicrobial resistance has five strategies as objectives: (1) antimicrobial resistance should be acknowledged and understood better; (2) Improvement of the knowledge by thorough research; (3) Limiting the occurrence of infection; (4) Adjustment of the antimicrobial agent usage; and (5) Consideration of other countries and major investment in the healthcare products.

Accessories which made a tremendous change in the history of mankind also pose a threat to cause diseases by acting as a reservoir of pathogenic micro-organisms. Enough attention should be given to the things which are unavoidable to man and research should be done to find the solution to the common pathogenic microbes present in the accessories. This review throws light on the accessories and gadgets which are very nearer and dearer to us and it's another side of causing health ailments due to the accumulation of micro-organisms. Furthermore, studies and grandness should be given for the things like earphones, key chains, zips of bags, handles of vehicles, etc. The studies should not be limited to the hospital environment and must be distributed to various environments like Banks, Schools, Colleges, and the places where more people gather like railway stations, bus stands, etc.

References

1. Adamu L, Edeghagba B, Olatomi F, Ezeokoli O, Elijah, A (2012) Microorganisms associated with commercial motorcycle helmets in Lagos metropolis. J Microbiol Biotechnol Food Sci (Slovak Republic)
2. Agarwal M, Hamilton-Stewart P, Dixon RA (2002) Contaminated operating room boots: the potential for infection. Am J Infect Control 30:179–183. https://doi.org/10.1067/mic.2002.119513
3. Al-Ahmad A, Wiedmann-Al-Ahmad M, Deimling D, Jaser C, Pelz K, Wittmer A, Ratka-Krüger P (2010) An antimicrobial effect from silver-coated toothbrush heads. Am J Dent 23:251–254
4. Alothman A, Bukhari A, Aljohani S, Muhanaa A (2009) Should we recommend stethoscope disinfection before daily usage as an infection control rule? Open Infect Dis J 3:80–82. https://doi.org/10.2174/1874279300903010080
5. Alrazeeni D, Al Suf MS (2014) Nosocomial infections in ambulances and efectiveness of ambulance fumigation techniques in Saudi Arabia: Phase I study. Saudi Med J 35:1354–1360
6. Balamurugan A, Balossier G, Laurent-Maquin D, Pina S, Rebelo AHS, Faure J, Ferreira JMF (2008) An in vitro biological and anti-bacterial study on a sol-gel derived silver-incorporated bioglass system. Dent Mater 24:1343–1351. https://doi.org/10.1016/j.dental.2008.02.015
7. Barillo DJ, Marx DE (2014) Silver in medicine: a brief history BC 335 to present. Burns 40:S3–S8. https://doi.org/10.1016/j.burns.2014.09.009
8. Brady RR, Fraser SF, Dunlop MG, Paterson-Brown S, Gibb AP (2007) Bacterial contamination of mobile communication devices in the operative environment. J Hosp Infect 66:397–398. https://doi.org/10.1016/j.jhin.2007.04.015
9. Bures S, Fishbain JT, Uyehara CFT, Parker JM, Berg BW (2000) Computer keyboards and faucet handles as reservoirs of nosocomial pathogens in the intensive care unit. Am J Infect Control 28:465–471. https://doi.org/10.1067/mic.2000.107267

10. Casey AL, Adams D, Karpanen TJ, Lambert PA, Cookson BD, Nightingale P, Miruszenko L, Shillam R, Christian P, Elliott TSJ (2010) Role of copper in reducing hospital environment contamination. J Hosp Infect 74:72–77. https://doi.org/10.1016/j.jhin.2009.08.018
11. Casey AL, Karpanen TJ, Adams D, Lambert PA, Nightingale P, Miruszenko L, Elliott TSJ (2011) A comparative study to evaluate surface microbial contamination associated with copper-containing and stainless steel pens used by nurses in the critical care unit. Am J Infect Control 39:e52–e54. https://doi.org/10.1016/j.ajic.2010.12.012
12. Cataño JC, Echeverri LM, Szela C (2012) Bacterial contamination of clothes and environmental items in a third-level hospital in Colombia. Interdisc Perspect Infect Dis. https://doi.org/10.1155/2012/507640
13. Chen W, Oh S, Ong AP, Oh N, Liu Y, Courtney HS, Appleford M, Ong JL (2007) Antibacterial and osteogenic properties of silver-containing hydroxyapatite coatings produced using a sol gel process. J Biomed Mater Res A 82:899–906. https://doi.org/10.1002/jbm.a.31197
14. Ciragil P, Gul M, Aral M (2006) Bacterial contamination of computers and telephones in a university hospital in Turkey. J Hosp Infect 62:247–248. https://doi.org/10.1016/j.jhin.2005.08.008
15. Codish S, Toledano R, Novack V, Sherf M, Borer A (2015) Effectiveness of stringent decontamination of computer input devices in the era of electronic medical records and bedside computing: a randomized controlled trial. Am J Infect Control 43:644–646. https://doi.org/10.1016/j.ajic.2014.09.011
16. D'Antonio NN, Rihs JD, Stout JE, Yu VL (2013) Computer keyboard covers impregnated with a novel antimicrobial polymer significantly reduce microbial contamination. Am J Infect Control 41:337–339. https://doi.org/10.1016/j.ajic.2012.03.030
17. Dasgupta S, Das S, Chawan NS, Hazra A (2015) Nosocomial infections in the intensive care unit: Incidence, risk factors, outcome and associated pathogens in a public tertiary teaching hospital of Eastern India. Indian J Crit Care Med 19:14–20. https://doi.org/10.4103/0972-5229.148633
18. Dhand A, Sakoulas G (2014) Daptomycin in combination with other antibiotics for the treatment of complicated Methicillin-resistant staphylococcus aureus bacteremia. Clin Ther 36:1303–1316. https://doi.org/10.1016/j.clinthera.2014.09.005
19. Doğan M, Feyzioğlu B, Ozdemir M, Baysal B (2008) Investigation of microbial colonization of computer keyboards used inside and outside hospital environments. Mikrobiyol Bul 42:331–336
20. Feldman J, Feldman J, Feldman M (2012) Women doctors' purses as an unrecognized fomite. Del Med J 84:277–280
21. Fraser MA, Girling SJ (2009) Bacterial carriage of computer keyboards in veterinary practices in Scotland. Veter Rec 165:26–27
22. Fukada T, Iwakiri H, Ozaki M (2008) Anaesthetists' role in computer keyboard contamination in an operating room. J Hosp Infect 70:148–153. https://doi.org/10.1016/j.jhin.2008.05.023
23. Grass G, Rensing C, Solioz M (2011) Metallic copper as an antimicrobial surface. Appl Environ Microbiol 77:1541–1547. https://doi.org/10.1128/AEM.02766-10
24. Gray TB, Cursons RT, Sherwan JF, Rose PR (1995) Acanthamoeba, bacterial, and fungal contamination of contact lens storage cases. Br J Ophthalmol 79:601–605. https://doi.org/10.1136/bjo.79.6.601
25. Hakuno H, Oie S, Furukawa H (2012) Contamination of wheelchairs by Staphylococcus aureus. J Hosp Admin 2. https://doi.org/10.5430/jha.v2n2p55
26. Halton K, Arora V, Singh V, Ghantoji SS, Shah DN, Garey KW (2011) Bacterial colonization on writing pens touched by healthcare professionals and hospitalized patients with and without cleaning the pen with alcohol-based hand sanitizing agent. Clin Microbiol Infect 17:868–869. https://doi.org/10.1111/j.1469-0691.2011.03494.x
27. Jamal MA, Rosenblatt J, Jiang Y, Hachem R, Chaftari A-M, Raad II (2014) Prevention of transmission of multidrug-resistant organisms during catheter exchange using antimicrobial catheters. Antimicrob Agents Chemother 58:5291–5296. https://doi.org/10.1128/AAC.02886-14

28. Jeske H-C, Tiefenthaler W, Hohlrieder M, Hinterberger G, Benzer A (2007) Bacterial contamination of anaesthetists' hands by personal mobile phone and fixed phone use in the operating theatre. Anaesthesia 62:904–906. https://doi.org/10.1111/j.1365-2044.2007.05172.x
29. Jones R, Hutton A, Mariyaselvam M, Hodges E, Wong K, Blunt M, Young P (2015) Keyboard cleanliness: a controlled study of the residual effect of chlorhexidine gluconate. Am J Infect Control 43:289–291. https://doi.org/10.1016/j.ajic.2014.12.002
30. Kaliyadan F, Aboulmagd E, Amin TT (2014) Antimicrobial activity of commercial "antibacterial" handwashes and soaps. Indian Dermatol Online J 5:344–346. https://doi.org/10.4103/2229-5178.137799
31. Kandel CE, Simor AE, Redelmeier DA (2014) Elevator buttons as unrecognized sources of bacterial colonization in hospitals. Open Med 8:e81–e86
32. Karibasappa GN, Nagesh L, Sujatha BK (2011) Assessment of microbial contamination of toothbrush head: an in vitro study. Indian J Dent Res 22:2–5. https://doi.org/10.4103/0970-9290.79965
33. Karpanen TJ, Casey AL, Lambert PA, Cookson BD, Nightingale P, Miruszenko L, Elliott TSJ (2012) The antimicrobial efficacy of copper alloy furnishing in the clinical environment: a crossover study. Infect Control Hosp Epidemiol 33:3–9. https://doi.org/10.1086/663644
34. Kilic IH, Ozaslan M, Karagoz ID, Zer Y, Davutoglu V (2009) The microbial colonisation of mobile phone used by healthcare staffs. Pak J Biol Sci 12:882–884. https://doi.org/10.3923/pjbs.2009.882.884
35. Kokare CR, Chakraborty S, Khopade AN, Mahadik KR (2009) Biofilm: Importance and applications. Indian J Biotechnol 8:159–168
36. Kumar A, Vemula PK, Ajayan PM, John G (2008) Silver-nanoparticle-embedded antimicrobial paints based on vegetable oil. Nat Mater 7:236–241. https://doi.org/10.1038/nmat2099
37. Kuo-Hu Chen L-RC (2014) Contamination of medical charts: an important source of potential infection in hospitals. PLoS One 9:e78512. https://doi.org/10.1371/journal.pone.0078512
38. Lebeaux D, Joly D, Zahar J-R (2014) Long-term intravenous catheter-related infections. Revue du Praticien 64:626–633
39. Lee P-C, Chong ETJ, Abu Nor S (2014) Evaluation of microbial contamination on contact lenses among university students. Trans Sci Technol 1:36–42
40. Malini M, Thomas T, Bhargava D, Girija S (2012) Microbiology of the white coat in a dental operatory. Indian J Dent Res 23:841. https://doi.org/10.4103/0970-9290.111289
41. Meadow JF, Altrichter AE, Green JL (2014) Mobile phones carry the personal microbiome of their owners. PeerJ. https://doi.org/10.7717/peerj.447
42. Mehta V, Singh M, Vaishya S, Gupta S (2006) Economics of head injuries. Neurol India 54:78. https://doi.org/10.4103/0028-3886.24714
43. Meier S, Collier C, Scaletta MG, Stephens J, Kimbrough R, Kettering JD (1996) An in vitro investigation of the efficacy of CPC for use in toothbrush decontamination. J Dental Hygiene: JDH/Am Dental Hygienists' Assoc 70:161–165
44. Morillo-García Á, Aldana-Espinal JM, Olry de Labry-Lima A, Valencia-Martín R, López-Márquez R, Loscertales-Abril M, Conde-Herrera M (2015) Hospital costs associated with nosocomial infections in a pediatric intensive care unit. Gac Sanit 29:282–287. https://doi.org/10.1016/j.gaceta.2015.02.008
45. Mukherjee K, Mathuriya S, Tripathi M, Tewari M (2014) Profile of patients with head injury among vehicular accidents: an experience from a tertiary care centre of India. Neurol India 62:610. https://doi.org/10.4103/0028-3886.149382
46. Nagajothi J, Jeyakumari D, Vigneshwaran S, Kumar RP, Bharatwaj RS, Bagyalakshmi R (2015) Study of prevalence of microbial contamination with its antibiotic resistance pattern in automated teller machine in and around Puducherry, India. J Earth Environ Health Sci 1:27
47. Neely AN, Maley MP, Warden GD, Seifert H, Wisplinghoff H (1999) Computer keyboards as reservoirs for Acinetobacter baumannii in a burn hospital. Clin Infect Dis 29:1358–1360
48. Nworie A., Department of Medical Laboratory Science, C. of H.S., Ayeni, J.A., Department of Medical Microbiology, F.S. of M.L.S., Eze, U.A., Department of Medical Laboratory Science, C. of H.S., Azi, S.O., Department of Medical Laboratory Science, C. of H.S. (2012) Bacterial

contamination of door handles/knobs in selected public conveniences in Abuja metropolis. A Public Health Threat, Nigeria
49. Page K, Correia A, Wilson M, Allan E, Parkin IP (2014) Light-activated antibacterial screen protectors for mobile telephones and tablet computers. J Photochem Photobiol, A 296:19–24. https://doi.org/10.1016/j.jphotochem.2014.08.011
50. Patel S, Porter K, Sammons RL (2010) Are computer keyboards a cross-infection risk in a dental clinic? J Infect Prev 11:206–211. https://doi.org/10.1177/1757177410384892
51. Pihl M, Davies JR, Johansson A-C, Svensäter G (2013) Bacteria on catheters in patients undergoing peritoneal dialysis. Perit Dial Int 33:51–59. https://doi.org/10.3747/pdi.2011.00320
52. Pisipati S, Bassett D, Pearce I (2009) Do neckties and pens act as vectors of hospital-acquired infections? BJU Int 103:1604–1605. https://doi.org/10.1111/j.1464-410X.2009.08440.x
53. Ramesh J, Carter AO, Campbell MH, Gibbons N, Powlett C, Moseley H Sr, Lewis D, Carter T (2008) Use of mobile phones by medical staff at Queen Elizabeth Hospital, Barbados: evidence for both benefit and harm. J Hosp Infect 70:160–165. https://doi.org/10.1016/j.jhin.2008.06.007
54. Roe D, Karandikar B, Bonn-Savage N, Gibbins B, Roullet J-B (2008) Antimicrobial surface functionalization of plastic catheters by silver nanoparticles. J Antimicrob Chemother 61:869–876. https://doi.org/10.1093/jac/dkn034
55. Rutala WA, White MS, Gergen MF, Weber DJ (2006) Bacterial contamination of keyboards: efficacy and functional impact of disinfectants. Infect Control Hosp Epidemiol 27:372–377. https://doi.org/10.1086/503340
56. Sato S, Pedrazzi V, Guimarães Lara EH, Panzeri H, De Albuquerque Jr, RF, Ito IY (2005) Antimicrobial spray for toothbrush disinfection: an in vivo evaluation. Quintessence Int 36:812–816
57. Schultz M, Gill J, Zubairi S, Huber R, Gordin F (2003) Bacterial contamination of computer keyboards in a teaching hospital. Infect Control Hosp Epidemiol 24:302–303. https://doi.org/10.1086/502200
58. Sedighi I, Alikhani MY, Ramezani S, Nazari M, Nejad ASM (2015) Bacterial contamination of mobile phones of health care providers in a teaching hospital in Hamadan Province, Iran. Arch Clin Infect Dis 10. https://doi.org/10.5812/archcid.10(2)2015.22104
59. Sepehri (2009) Bacterial contamination and resistance to commonly used antimicrobials of healthcare workers' mobile phones in teaching hospitals, Kerman, Iran. Am J Appl Sci 6:806–810. https://doi.org/10.3844/ajassp.2009.806.810
60. Shalini S, Vidyasree MD, Abiselvi A, Gopalakrishnan S (2015) Impact and effect of nosocomial infections: a review. Res J Pharm, Biol Chem Sci 6:947–951
61. Sharp K, Haysom IW (2003) The survival and recovery of bacteria in vacuum cleaner dust. J Royal Soc Promot Health 123:39–45. https://doi.org/10.1177/146642400312300114
62. Shiferaw T, Beyene G, Kassa T, Sewunet T (2013) Bacterial contamination, bacterial profile and antimicrobial susceptibility pattern of isolates from stethoscopes at Jimma University specialized hospital. Ann Clin Microbiol Antimicrobials 12. https://doi.org/10.1186/1476-0711-12-39
63. Singh V, Chauhan PK, Kanta R, Puri A, Saini JS (2011) Are we aware how contaminated our mobile phones with pathogenic bacteria. Int J Pharm Pharm Sci 3:275–276
64. Splete H (2012) Blognosis: will use of EMRs fuel keyboard contamination? Oncology Report 23
65. Szczotka-Flynn LB, Pearlman E, Ghannoum M (2010) Microbial contamination of contact lenses, lens care solutions, and their accessories: a literature review. Eye Contact Lens 36:116–129. https://doi.org/10.1097/ICL.0b013e3181d20cae
66. Tadiparthi S, Shokrollahi K, Juma A, Croall J (2007) Using marker pens on patients: a potential source of cross infection with MRSA. Ann R Coll Surg Engl 89:661–664. https://doi.org/10.1308/003588407X209419
67. Teng S-O, Lee W-S, Ou T-Y, Hsieh Y-C, Lee W-C, Lin Y-C (2009) Bacterial contamination of patients' medical charts in a surgical ward and the intensive care unit: impact on nosocomial infections. J Microbiol Immunol Infect 42:86–91

68. Ting CY, Cheng DR, Kotsanas D, Kirkwood CD, Bogdanovic-Sakran N, Stuart RL, Buttery JP (2015) Are identity badges and lanyards in pediatrics potentially contaminated with viral pathogens? Am J Infect Control. https://doi.org/10.1016/j.ajic.2015.06.024
69. Treakle AM, Thom KA, Furuno JP, Strauss SM, Harris AD, Perencevich EN (2009) Bacterial contamination of health care workers' white coats. Am J Infect Control 37:101–105. https://doi.org/10.1016/j.ajic.2008.03.009
70. Ulger F, Esen S, Dilek A, Yanik K, Gunaydin M, Leblebicioglu H (2009) Are we aware how contaminated our mobile phones with nosocomial pathogens? Ann Clin Microbiol Antimicrobials 8. https://doi.org/10.1186/1476-0711-8-7
71. Uneke CJ, Ogbonna A, Oyibo PG, Onu CM (2010) Bacterial contamination of stethoscopes used by health workers: public health implications. J Infect Dev Ctries 4:436–441
72. Ustun C, Cihangiroglu M (2012) Health care workers' mobile phones: a potential cause of microbial cross-contamination between hospitals and community. J Occup Environ Hyg 9:538–542. https://doi.org/10.1080/15459624.2012.697419
73. Venkatesh AK, Lankford MG, Rooney DM, Blachford T, Watts CM, Noskin GA (2008) Use of electronic alerts to enhance hand hygiene compliance and decrease transmission of vancomycin-resistant Enterococcus in a hematology unit. Am J Infect Control 36:199–205. https://doi.org/10.1016/j.ajic.2007.11.005
74. Walia SS, Manchanda A, Narang RS, Anup N, Singh, B., Kahlon SS (2014) Cellular telephone as reservoir of bacterial contamination: Myth or Fact. J Clin Diagn Res 8:50–53. https://doi.org/10.7860/JCDR/2014/6398.3948
75. Wilson APR, Ostro P, Magnussen M, Cooper B, Keyboard Study Group (2008) Laboratory and in-use assessment of methicillin-resistant Staphylococcus aureus contamination of ergonomic computer keyboards for ward use. Am J Infect Control 36:e19–25. https://doi.org/10.1016/j.ajic.2008.09.001
76. Wong D, Nye K, Hollis P (1991) Microbial flora on doctors' white coats. BMJ 303:1602–1604
77. Yildirim I, Ceyhan M, Cengiz AB, Bagdat A, Barin C, Kutluk T, Gur D (2008) A prospective comparative study of the relationship between different types of ring and microbial hand colonization among pediatric intensive care unit nurses. Int J Nurs Stud 45:1572–1576. https://doi.org/10.1016/j.ijnurstu.2008.02.010
78. Young JM, Naqvi M, Richards L (2005) Microbial contamination of hospital bed handsets. Am J Infect Control 33:170–174. https://doi.org/10.1016/j.ajic.2004.11.005

Utilization of Waste Tires Toward Concrete Production and Decomposition of Tires by Pyrolysis

J. Karthikeyan, K. J. Rupesh, A. Arumugam, and S. Sudalai

Abstract The increasing population and quest for comfortness mandate economic development. The transportation sector plays a major role in connecting people and materials in all possible ways. The tires are the most common bulk waste generated from the transport sector. Research has been conducted on greening the production of tires and extending their durability of tires. The disposal of used tires grabs the attention of waste managers of any administration due to its characteristics. The waste tires in the dumpsites fuels fire accidents and the combustion of waste tires invite air pollution problems and the disposal of waste tires in water bodies leads to serious pollution issues. The process of waste tire utilization includes collection, shredding, steel liberation, cleaning, packing, and transportation to the processing unit. The material and energy recovery from the management of end-of-life tires encourages waste managers since the revenue generated from the resources will support the overall process. The end-of-life tire scraps can be used for various upcycling applications but establishing the local supply chain for the products from waste management encourages sustainability. The present article discusses the utilization of waste tires in construction and pyrolysis.

Keywords Waste tire · Pyrolysis · Recycled · Construction material

1 Introduction

From the world bank report for the year 2018, global annual waste generation was expected to grow 70% in the next 30 years which is about 3.4 billion tons per year. The generation of waste in the year 2016 was 2.01 billion tons. The generation of waste

J. Karthikeyan · K. J. Rupesh · S. Sudalai (✉)
Centre for Pollution Control and Environmental Engineering, School of Engineering and Technology, Pondicherry University, Kalapet, Puducherry 605014, India
e-mail: ssudalai.cpe@gmail.com

K. J. Rupesh · A. Arumugam
Bioprocess Intensification Laboratory, Centre for Bioenergy, School of Chemical and Biotechnology, SASTRA Deemed University, Thirumalaisamudram, Thanjavur, Tamil Nadu, India

N. A. Siddiqui et al. (eds.), *Advances in Waste Management*, Lecture Notes in Civil Engineering 301, https://doi.org/10.1007/978-981-19-7506-6_6

not only contributes the space occupancy, but it will also affect both environment and living organisms. So recycling is one of the important processes to reduce the generation of waste materials [2]. There are one billion end-of-life tires produced annually worldwide and only <50% are recycled remaining things are in landfills. Landfills pose dangers of accidental fires and consume the valued space also leaches of toxic substances into the environment. So all government sectors focused on encouraged to recycle tires to improve the utilization rate of tires [4]. Retreading of end-life tires, manufacturing of rubber molded products, production of carbon black and oil/gas by pyrolysis, and reinforcing of rubber particles into different types of concretes are the most common methods used for reducing the number of waste tires [19].

The generation of waste tires is more than 0.44 million tons in the UK. Around 21% of waste tires are sent for shredding and used as a raw material for other processes, 22% of waste tires are used for energy recovery, and 34% of waste tires are disposed of in landfills. It is also very difficult to recover tires when mixed with other types of solid waste [24].

Around 51 million passenger tires reach their end life and only 9% of them was recycled locally only in Australia [38]. The recycled tire crumb in concrete results in improvement of elasticity, abrasion resistance, and damping property. The unit weight of the concrete was decreased addition of tire crumb rubber [26]. Because tire crumb rubber has lower specific gravity. Used tires have a high potential for recycling compared to various recycled materials. Based on information in 1994, Europe produces two million scrap tires per year and 46% of tires are landfilled and 31% of tires are used for energy recovery. Tire-derived fuel (TDF) is one of the important recycling options used in pulp, paper mills, cement kilns power stations, and industrial boilers. The energy content is about 32.0 GJ/ton and CO_2 emission is about 2270 kg/ton. If the tire particles were mixed with sand under different proportions like 10–40%, the shear strength and angle of internal friction were improved [25, 31].

2 Waste Tire Generation

Annually 1100 crore tons of solid waste are generated globally. Taiwan produced around 1 lakh tons of waste tires annually, in the US, 27 crores of waste tires are generated annually and 11 crore waste tires are generated in Japan [43]. Statistics show that each person generates one ton of waste annually (Fig. 1). Compared to the year 2000 amount of waste generated will be doubled in 2025. The solid wastes are classified into three types, (i) Municipal, (ii) Commercial, and industrial (iii) Construction and demolition. The tires are under commercial and industrial solid waste and 75% of end-of-life tires reach landfills. The landfilling of solid waste causes environmental and health hazards due to release of toxics, leachates, and greenhouse gases. Landfilling is the last option for non-recyclable waste because the wastes are recycled for various applications the remaining things are only going to landfills [16].

Fig. 1 Tire waste yard

3 Environmental Pollution by Waste Tires

Annually one billion tires are discarded, and only a fraction of tires get recycled remaining are placed in a storage yard and used in landfills. Landfilling requires a larger area and some of the tires are burnt off resulting in the release of toxic fumes like carbon monoxide, styrene, butadiene, and polyaromatic hydrocarbons which threatens human health [29]. The decomposition of the tiring process also affects the environment. During the pyrolysis process, carbon monoxide and particulate matter are released at their higher concentrations which results in a harmful effect on the environment [41]. The wastewater from the pyrolysis process is also contaminated highly with a higher amount of TDS, TSS, COD, chloride, and heavy metals like copper and manganese. The waste from the pyrolysis process is directly fed into a sewage system without any treatment [3].

By static material flow analysis, 80% of the tire rubber remained in use, while 14% of the tires are re-treaded, incinerate, or recycled, and 6% were emitted as tire wear particles to air soil or surface water. The range of dissipated TWPs is 21200 tons among them 6% were microscale (size range between 0.1 and 10 μm), and 0.3% were nanoscale (size is below 0.1 μm). From the mass balance on the substance, the level shows tire wear particles containing the emission of carbon black in the form of airborne or surface water or soil. The range of carbon black emission is 5500 tons/year. About 3600 tons/year were no exhaust emission (tire, road- surface wear) regarding the air pollution from road vehicles [33].

Leaching of waste tires on groundwater and soil is in civil engineering. Butanone and toluene are emitted from rubber fragments. A high concentration of benzene also leached from tire chips in range of 0.0115 mg/L and toluene range is 0.0112 mg/L. From various research, metals like aniline and phenolic compounds like cresol

Table 1 Contribution of microplastics by TWPs in %

Country	Contribution of microplastics by TWPs in %
Germany	24–30
Switzerland	94
China	54
Denmark	56
Sweden	61–79

isomers are also leached from shredded tire. Aniline can reach a concentration of 5.73 mg/kg. Diethyl phthalate has a lower concentration value (3 μg/kg). And hydrocarbon concentration is about 2.3 mg/L [29].

The emission of tire wear particles contributes to the emission of microplastics according to the different studies from various countries [30] (Table 1).

According to the predictions, the number of waste tire generation increased with 1200 million tires per year until 2030. Dumped and stocked waste tires are at a risk of burning. In 2016, in Serena waste tires caught fire and many people were evacuated. Tire fire results in the emission of hazardous compounds and fine particulates which results in carcinogenic effects on humans [21]. Estimation showed that around 70,000 to 90,000 tons of tires for many years in Serena [18].

Ribbed smoked sheets (RSS) are the main rubber product which are used in the production of tires. Thailand researchers determined emissions during the manufacturing of RSS by a life cycle inventory methodology. From the results, in Japan tire production contributes to 90% of the poly aromatic hydrocarbons and 90% of the SO_x emissions. 60% of particulate matter and 80% of the CO_2 are released during the drying process without any pollution control devices [32].

4 Classification of Waste Tires

The waste tires are classified under the commission europeenne de normalization (CEN) (European standards committee) workshop agreement, ASTM D6270-08 and ASTM D5681-18. By ASTM standards, the waste tire rubbers are classified into granulated, powdered, tire chips, tire-derived aggregate rubber, and rough shreds. Waste tire rubber can be further classified based on the size, crumb rubber, ground rubber, fiber rubber aggregate, rubber chips, chipped or shredded rubber aggregate, and the ground rubber and crumb rubber. The shape of the waste tire rubbers is non-spherical and some of them have a basic geometrical shape [29] (Table 2).

Table 2 Classification of tire rubber

Types of tire rubber	Size in mm
Chipped or shredded rubber	13–76
Crumb rubber	0.425–4.75
Ground rubber	<0.425
Fiber rubber aggregate	12.5
Rubber chips	10–25
Granulated rubber	0.425–12
Carbon products	<500 μm
Pyrolitic char	<10

5 Utilization of Tire Rubbers as a Construction Material

The world is shifting on the way to utilization of material completely and recyclable condition. It focuses on the use of the materials for the longest time possible. By this aspect, the tires are classified under different categories and they are used for different applications like chipped rubber, tire crumb rubbers reinforced with concretes, and the ground tire rubbers are used in the manufacture of ground covers in the playground, walkway tiles, running track, conveyor belts, shoes, car mats7. The utilization of tires in various applications plays a vital role, Because only 10% of tires are recycled globally, and the remaining are under landfilling which results in environmental problems [16].

Two processes are usually used for using tire in asphalt pavement, dry process, and wet process. Crumb rubber and bitumen are mixed for 45 min at high temperature for wet process. The range of crumb rubber is 18–22% of the bitumen weight, by this reaction, the bitumen will harden and the properties also increase. Ground rubber can be used in place of fine aggregates in asphalt pavement with a replacement rate of 1–2%. Tire rubbers were used in a road foundation because of their lightweight, higher hydraulic conductivity, lower thermal conductivity, and higher shear strength at large strain compared to the compacted soil [24].

5.1 Waste Tire Rubber in Self-compacting Concrete

The self-compacting concrete is environmentally friendly due to the addition of superplasticizers into fresh mixtures that helps avoiding the vibration/compaction of the concrete. Here the rubber powder was filtered through 30 and 50 mm sieves and then reinforced with concrete at different volume ratios ranging from 5 to 20%, respectively. Results show that the addition of 5% of waste tire rubber increased compressive strength and anti-sulfate corrosion. The shrinkage was higher and low ultrasonic pulse velocity that observed additional powder was included [43].

5.2 *Steel Fiber Reinforced Concrete*

Steel cords extracted from a un vulcanized rubber can be used in concrete9. The length of steel fiber ranges from 25 to 40 mm and the dosage of fiber is 30–45 kg/m^3. Both manufactured and recycled steel fiber are reinforced at the same dosage. As a result, the RTSC (Recycled tire steel cords) have better post-cracking strength and flexural strength [23]. The recycled steel fiber ARE reinforced with concrete with different dosages rates of 15, 30, 60 kg/m^3. Then various splitting test and 45 axial compression test was done. The results fulfill the requirements specified by building codes NSR-10 and ACI-318 [10].

5.3 *Recycled Tire Crumb in Lightweight Cellular Concrete (LCC)*

The use of tire crumbs in construction materials will reduce environmental issues and also be economically beneficial. Tire crumb is used as a filler material in lightweight cellular concrete [6]. Adding the tire products into the concrete will increase insulation properties. Crumb rubber when surface treated using NaOH solution shows better compressive strength. The results were confirmed using SEM and EDS analysis.

6 Properties of Rubber-Based Concrete

After the reinforcement of different types of rubbers into the concrete, the different physical and chemical properties are measured [8]. The density, workability, and water absorption under physical properties and flexural strength, compressive strength, impact strength, and abrasion resistance, under mechanical properties are tested [20]. The property values are not the same for every type of concrete [39]. Some important properties are described here.

6.1 *Workability*

Workability is the most important property. Due to the reinforcement of rubber particles into concrete, the workability may have been affected. Due to the highly hydrophobic nature of the rubber workability property was decreased. The superplasticizer also affects the workability of the concrete. The addition of cryogenic ground tire rubber shows better workability compared to mechanically ground tire rubber [1].

6.2 Density

Rubber has low density, so the concrete density was also decreased compared to normal concrete. The density of the particles depends upon the specific gravity. The specific gravity of the tire varies between 0.6 and 1.15 m^3/kg. It is significantly lower than the conventional aggregate. The specific gravity of aggregate will be around 2.65 m^3/kg [37]. Research shows that the rubberized concrete mixtures have lower density, high ductility, very high toughness [43].

6.3 Compressive Strength

Compressive strength is one of the very important properties of the concrete and it will differ depending on the amount of rubber content and steel fibers present on the particle. The majority of studies show that if the rubber content increase, the compressive strength will decrease, because rubbers have poor bonding behavior and soft material property. Also if the porosity is low it will increase the compressive strength.

7 Pyrolysis Process

Pyrolysis or thermal cracking is the thermal decomposition of organic substances under lack of air with or without the aid of a catalyst. The organic compounds breakdown at very high temperature of about 300–1300 °C. Pyrolysis process has several advantages, pyrolysis can convert useless, waste materials into high calorific valued fuels. Steps for recycling waste tires are heating whole or halved or shredded tires in absence of air [11]. The rubber will soften and will disintegrate into smaller molecules which will eventually vaporize. These vapors are collected as gas or condensed into an liquid which is called pyrolysis oil. The pyrolysis oil can be used as an substitute to fossil fuels. Pyrolysis oil is usually pretreating before being used in machines (Fig. 3). Pyrolysis when performed well at optimal operating conditions will release almost negligible amount of gases into the atmosphere and is environmentally friendly [14, 15, 34].The process of pyrolysis is classified into three types (i) slow pyrolysis, (ii) fast pyrolysis, and (iii) catalytic pyrolysis. In slow pyrolysis here the tire was decomposed under lower temperatures prolonged time [28]. This process is particularly used for char production, even though small amount of liquid and gas fractions are produced. In fast pyrolysis, the tire was decomposed under high temperatures to achieve rapid decomposition [30]. The fast pyrolysis process is considered to be the most effective means of converting waste tires into high-calorific liquid fuel with very high conversion [42]. Nowadays, fast pyrolysis processes are assisted by microwave operations

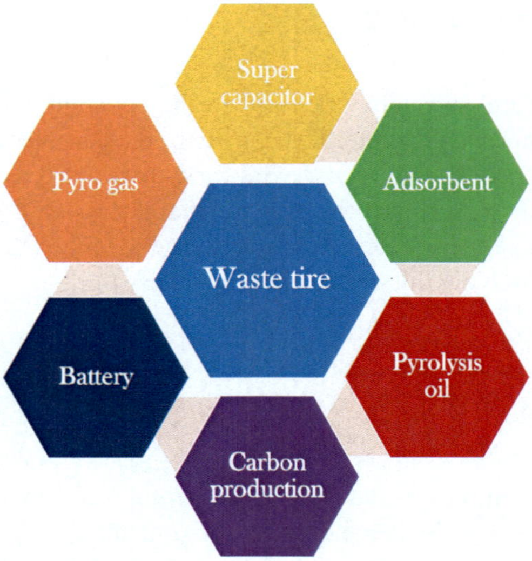

Fig. 3 Waste tire pyrolysis products and their application

or thermal plasma assisted. Catalytic pyrolysis contains catalysts to improve product conversion rate.

Table 3 reports the literature analysis of pyrolysis of waste materials.

8 Conclusion

The various results show that the mechanical, physical, and chemical properties of the different types of concrete depend on the nature of concrete and also rubber particles present in it. An increase in rubber particles reduces the mechanical properties due to low viscosity, and ductility. The recycling of waste tire rubber into various applications is very useful to make a clean environment. Increasing amount of tire disposal each year and increasing energy need make pyrolysis process and extraction of other useful products from tires inevitable. Also health and environmental considerations also play an important role to find a proper decomposition method for waste tires.

Table 3 Literature studies for the pyrolysis process

References	Type of material used	Type of process	Process parameters	Measured properties	Results
[36]	Agricultural and industrial waste	Pyrolysis	Temperature range 300–700	Physio-chemical properties	Recovery of Fe and Mn
[15]	Sewage sludge and waste TYREs	Co-pyrolysis	Temperature range 300–700	Adsorption of cd and tc	–
[35]	Agricultural and industrial waste	Co-pyrolysis	Temperature range 300–700°	Stability, electrical conductivity	Economically and environmentally benefit
[12]	Waste tire pyrolysis	–	–	Selection of process parameters to reduce pollutants	–
[22]	–	The effluent treatment plant for pyrolysis	–	Design parameters	–
[13]	Crumb tire rubber	Pyrolysis	–	Rheological properties	–
[5]	Scrap tire	Pyrolysis	Temperature range:400° at fixed bed batch pyrolysis	Viscosity, heating value	Produced pyro oil used for furnace oil in boiler operation

References

1. Adeboje AO, Kupolati WK, Sadiku ER, Ndambuki JM (2020) (2020) Characterization of modified crumb rubber concrete. Int J Sustain Dev Planning 15(3):377–383. https://doi.org/10.18280/ijsdp.150315
2. Alfayez S, Suleiman A, Nehdi M (2020) Recycling tire rubber in asphalt pavements: state of the art. Sustainability 12(21). https://doi.org/10.3390/su12219076
3. Arias AH, Alfonso MB, Girones l, Piccolo MC, Marcovecchio JE (2022) Synthetic microfibers and tyre wear particles pollution in aquatic systems: relevance and mitigation strategies. Environ Pollut 295:118607. https://doi.org/10.1016/j.envpol.2021.118607
4. Azevedo F, Pacheco-torgal F, Jesus C, de Aguiar J, Camoes A (2012) Properties and durability of HPC with tyre rubber wastes. Constr Build Mater 34:186–191. https://doi.org/10.1016/j.conbuildmat.2012.02.062
5. Aziz M, Rahman M, Molla H (2018) Design, fabrication and performance test of a fixed bed batch type pyrolysis plant with scrap tire in Bangladesh. J Radiat Res Appl Sci 11(4):311–316. https://doi.org/10.1016/j.jrras.2018.05.001
6. Baricevic A, Pezer M, Rukavina M, Serdar M, Stirmer N (2018) Effect of polymer fibers recycled from waste tires on properties of wet-sprayed concrete. Constr Build Mater 176:135–144. https://doi.org/10.1016/j.conbuildmat.2018.04.229
7. Baumgardner G, Hardee J, Negulescu I Williams E, Howard I, St John R (2014) Quantitative analysis of functional polymer in recycled tyre rubber used in modified asphalt binders. Road Mater Pav Design 15:263–278. https://doi.org/10.1080/14680629.2014.927413

8. Buss A, Kovaleski J, Pagani R, da silva V, Silva J (2019) Proposal to reuse rubber waste from end-of-life tires using thermosetting resin. Sustainability 11(24). https://doi.org/10.3390/su11246997
9. Caggiano A, Folino P, Lima C, Martinelli E, Pepe M (2017) On the mechanical response of hybrid fiber reinforced concrete with recycled and industrial steel fibers. Const Build Mater 147:286–295. https://doi.org/10.1016/j.conbuildmat.2017.04.160
10. Carrillo J, Lizarazo-marriaga J, Lamus F (2020) Properties of steel fiber reinforced concrete using either industrial or recycled fibers from waste tires. Fib Polym 21(9):2055–2067. https://doi.org/10.1007/s12221-020-1076-1
11. Centonze G, Leone M, Aiello M (2012) Steel fibers from waste tires as reinforcement in concrete: a mechanical characterization. Constr Build Mater 36:46–57. https://doi.org/10.1016/j.conbuildmat.2012.04.088
12. Chen G, Farooq M, Sun B, Lin F, Yan B, Rajput G, Chawla M (2021) Pollutants formation, distribution, and reaction mechanism during WT pyrolysis: a review. J Anal Appl Pyrolysis 157. https://doi.org/10.1016/j.jaap.2021.105218
13. Dong R, Zhao M, Tang N (2019) Characterization of crumb tire rubber lightly pyrolyzed in waste cooking oil and the properties of its modified bitumen. Constr Build Mater 195:10–18. https://doi.org/10.1016/j.conbuildmat.2018.11.044
14. Dong Y, Zhao Y, Hossain Md. U, He Y, Liu P (2021) Life cycle assessment of vehicle tires: a systematic review. Clean Environ Syst 2:100033. https://doi.org/10.1016/j.cesys.2021.100033
15. Fan X, Zhang J, Xie Y, Xu D, Liu Y, Liu J, Hou J (2021) Biochar produced from the co-pyrolysis of sewage sludge and waste tires for cadmium and tetracycline adsorption from water. Water Sci Technol 83(6):1429–1445. https://doi.org/10.2166/wst.2021.058
16. Ferdous W, Manalo A, Siddique R, Mendis P, Yan Z, Wong H, Lokuge W, Aravinthan T, Schubel P (2021). Recycling of landfill wastes (tyres, plastics and glass) in construction—a review on global waste generation, performance, application and future opportunities. Resourc Conserv Recycl 173. https://doi.org/10.1016/j.resconrec.2021.105745
17. Ferdous W, Manalo A, Siddique R, Mendis P, Zhuge Y, Wong HS, Lokuge W, Aravinthan T, Schubel P (2021) Recycling of landfill wastes (tyres, plastics and glass) in construction—a review on global waste generation, performance, application and future opportunities. Resourc Conserv Recyc 173:105745. https://doi.org/10.1016/j.resconrec.2021.105745
18. Formela K (2021) Sustainable development of waste tires recycling technologies—recent advances, challenges and future trends. Spec Iss: Recyc Neat Polym 4(3):209–222. https://doi.org/10.1016/j.aiepr.2021.06.004
19. Ge D, Yan K, You Z, Xu H (2016) Modification mechanism of asphalt binder with waste tire rubber and recycled polyethylene. Constr Build Mater 126:66–76. https://doi.org/10.1016/j.conbuildmat.2016.09.014
20. Ghadr S, Mirsalehi S, Assadi-langroudi A (2019) Compacted expansive elastic silt and tyre powder waste. Geomech Eng 18(5):535–543. https://doi.org/10.12989/gae.2019.18.5.535
21. Gupta T, Sharma R, Chaudhary S (2015) Influence of waste tyre fibers on strength, abrasion resistance and carbonation of concrete. Scientia Iranica 22(4):1481–1489
22. Hemanthnaidu A, Soni A, Khare A (2020) Design of effluent treatment plant for waste tyre pyrolysis industry, siltara, Raipur. J Indian Chem Soc 97(7):1043–1054
23. Hu H, Papastergiou P, Angelakopoulos H, Guadagnini M, Pilakoutas K (2018) Mechanical properties of SFRC using blended recycled tyre steel cords (RTSC) and recycled tyre steel fibres (RTSF). Const Build Mater 187:553–564. https://doi.org/10.1016/j.conbuildmat.2018.07.206
24. Huang Y, Bird R, Heidrich O (2007) A review of the use of recycled solid waste materials in asphalt pavements. Resourc Conserv Recyc 52(1):58–73. https://doi.org/10.1016/j.resconrec.2007.02.002
25. Imteaz M, Arulrajah A, Horpibulsuk S, Ahsan A (2018) Environmental suitability and carbon footprint savings of recycled tyre crumbs for road applications. Int J Environ Res 12(5):693–702. https://doi.org/10.1007/s41742-018-0126-7

26. Kashani A, Ngo T, Hemachandra P, Hajimohammadi A (2018) Effects of surface treatments of recycled tyre crumb on cement-rubber bonding in concrete composite foam. Constr Build Mater 171:467–473. https://doi.org/10.1016/j.conbuildmat.2018.03.163
27. Kashani A, Ngo T, Mendis P, Black J, Hajimohammadi A (2017b) A sustainable application of recycled tyre crumbs as insulator in lightweight cellular concrete. J Clean Prod 149:925–935. https://doi.org/10.1016/j.jclepro.2017.02.154
28. Kazmi S, Munir M, Wu Y (2021) Application of waste tire rubber and recycled aggregates in concrete products: a new compression casting approach. Resourc Conserv Recyc 167. https://doi.org/10.1016/j.resconrec.2020.105353
29. Li L, Yang J, Xiao H, Zhang L, Hu Z, Liu Y (2020b) Behavior of tire-geogrid-reinforced retaining wall system under dynamic vehicle load. Int J Geomech 20(4). https://doi.org/10.1061/(asce)gm.1943-5622.0001566
30. Mennekes D, Nowack B (2022) Tire wear particle emissions: measurement data where are you? Sci Total Environ 830:154655. https://doi.org/10.1016/j.scitotenv.2022.154655
31. Petrella A, di Mundo R, De gisi S, Todaro F, Labianca C, Notarnicola M (2019) Environmentally sustainable cement composites based on end-of-life tyre rubber and recycled waste porous glass. Materials 12(20). https://doi.org/10.3390/ma12203289
32. Phairuang W, Tekasakul P, Hata M, Tekasakul S, Chomanee J, Otani Y, Furuuchi M (2019) Estimation of air pollution from ribbed smoked sheet rubber in Thailand exports to japan as a pre-product of tires. Atmos Pollut Res 10(2):642–650. https://doi.org/10.1016/j.apr.2018.11.010
33. Prenner S, Allesch A, Staudner M, Rexeis M, Schwingshackl M, Huber-humer M, Part F (2021) Environ Pollut 290:118102. https://doi.org/10.1016/j.envpol.2021.118102
34. Rodriguez J, Lustosa J, Melo L, de Assis I, de Oliveira T (2021a) Co-pyrolysis of agricultural and industrial wastes changes the composition and stability of biochars and can improve their agricultural and environmental benefits. J Anal Appl Pyrolysis 155. https://doi.org/10.1016/j.jaap.2021.105036
35. Rodriguez J, Lustosa J, Melo L, de Assis I, de Oliveira T (2021b) Co-pyrolysis of agricultural and industrial wastes changes the composition and stability of biochars and can improve their agricultural and environmental benefits. J Anal Appl Pyrolysis 155. https://doi.org/10.1016/j.jaap.2021.105036
36. Rodriguez J, Lustosa J, Melo L, de Assis I, de Oliveira T (2022) Classifying the potential of biochars from agricultural and industrial waste for the recovery of Fe and Mg mining tailings. J Anal Appl Pyrolysis 161. https://doi.org/10.1016/j.jaap.2021.105383
37. Roychand R, Gravina R, Zhuge Y, Ma X, Youssf O, Mills J (2020) A comprehensive review on the mechanical properties of waste tire rubber concrete. Constr Build Mater 237. https://doi.org/10.1016/j.conbuildmat.2019.117651
38. Saberian M, Li J, Perera S, Ren G, Roychand R, Tokhi H (2020) An experimental study on the shear behaviour of recycled concrete aggregate incorporating recycled tyre waste. Constr Build Mater 264. https://doi.org/10.1016/j.conbuildmat.2020.120266
39. Sofi A (2018) Effect of waste tyre rubber on mechanical and durability properties of concrete—a review. Ain Shams Eng J 9(4):2691–2700. https://doi.org/10.1016/j.asej.2017.08.007
40. Wik A, Dave G (2009) Occurrence and effects of tire wear particles in the environment—a critical review and an initial risk assessment. Environ Pollut 157(1):1–11. https://doi.org/10.1016/j.envpol.2008.09.028
41. Yasar A, Rana S, Moniruzzaman M, Nazar M, Tabinda AB, Haider R, Ahmad A, Mukhtar A, Qyyum MA, Ullah S (2021) Quality and environmental impacts of oil production through pyrolysis of waste tyres. Environ Technol Innov 23:101565. https://doi.org/10.1016/j.eti.2021.101565
42. Yu H, Chen Y, Wu Q, Zhang L, Zhang Z, Zhang J, Miljkovic M, Oeser M (2020) Decision support for selecting optimal method of recycling waste tire rubber into wax—based warm mix asphalt based on fuzzy comprehensive evaluation. J Clean Prod 265. https://doi.org/10.1016/j.jclepro.2020.121781

43. Yung W, Yung l, Hua l (2013) A study of the durability properties of waste tire rubber applied to self-compacting concrete. Constr Build Mater 41:665–672. https://doi.org/10.1016/j.conbuildmat.2012.11.019

Comparative Analysis of Ultrafiltration for Produced Water from Oil and Gas Industry

Damini Rana, Sharika Bichinepally, Swati Kalra, Abhishek Nandan, and Nihal A. Siddiqui

Abstract Traditional technique of handling produced water from upstream oil and gas operations include reinjection for pressure support or deep well disposal. Studies into other water treatment methods will expand the end-use options of treated water in irrigation, water supply, or surface/groundwater discharge. In order to treat water to the extent required by each end-use option, most wastewater treatments include membrane separation to remove contaminants. Ultrafiltration helps to filter large contaminants from this water. Process economics of these membranes determines the feasibility of their application in large-scale industrial use depending on efficiency and costs. Critical analysis of the available literature on these membranes highlighted areas of potential future research that includes material characteristics in varying operating parameters (temperature, influent quality, etc.) This paper highlights the treatment efficiency of produced water using different membranes. According to the data obtained from these experiments, the best membrane for the treatment of produced water is zirconium oxide.

Keywords Ultra filtration · Membrane · Contaminants · Wastewater

1 Introduction

Produced water is a controversial topic amongst public as there is no concrete proof to say that this source of water is safe. Every year billions of barrels of the water are produced during extraction of oil and gas [16]. There is a heavy price tag that

D. Rana (✉)
Department of Environmental Sciences, Gurukula Kangri (Deemed to be University), Haridwar, Uttarakhand, India
e-mail: daminiranaofficial@gmail.com

Central Laboratory, Uttarakhand Pollution Control Board, Dehradun, India

S. Bichinepally · A. Nandan · N. A. Siddiqui
School of Engineering, UPES, Dehradun, Uttarakhand, India

S. Kalra
IIFM, Bhopal, Madhya Pradesh, India

N. A. Siddiqui et al. (eds.), *Advances in Waste Management*, Lecture Notes in Civil Engineering 301, https://doi.org/10.1007/978-981-19-7506-6_7

comes with managing produced water, which is directly related to the profitability of the production. Costs include the construction of treatment and disposal facilities, operational costs, chemical additives, permitting, transportation and reporting [19, 35]. These costs can easily exceed the value of the hydrocarbons produced and once this occurs the well is no longer worth producing [8, 28].

Produced water will have varying degrees of contaminates depending on the region from which the produced water was released from, and specifically the composition in the rock formation [13]. For example, the Bakken formation will produce water containing salinity levels 10 times that of seawater. Other contaminants found in produced wastewater include oil residue, grease, Naturally Occurring Radioactive Materials (NORMs) and inorganic/organic chemicals [7, 38]. In fact, there are over 165,000 measurements of the types of produced water compositions in the US alone, all of which may alter depending on the production rate (Oil and Gas Facilities 2012). Thus, there are numerous ways in which produced water may be treated. An examination of membrane separation will be conducted.

Quality of water issues within oil and gas industry vary in range depending on the oil and gas extraction technology used (whether it be conventional or unconventional method), the process in which the water is used for (such as Enhanced Recovery Oil Recovery processes) and the region in which the water is extracted from [12].

Depending on these factors, the water processed could have varying degrees of contaminants such as high salinity, hardness, various total suspended solids (TSS), volatile solids, and organic and inorganic matter [20]. These contaminants may be treated at various wastewater treatment plants—each treatment process may be configured specifically for the type of wastewater [36]. For example, the water produced from a residential community will not have the same content as the water received from a refinery [10]. Therefore, depending on the influent, the treatment process may be tailored to better treat the water [37].

For instance, the oil sands in Alberta, use about 8% of Alberta's water allocations while 2% is allocated to conventional oil and natural gas [27]. This water is extracted from the 1500 km long Athabasca River running from Alberta to the Columbia Glacier in Jasper National Park [27]. There are several processes for upstream oil and gas industry that requires water. Water is required for refining and processing oil and gas, electricity production in natural gas plants, and is a key component in drilling and hydraulically fractured wells [21, 34]. The processed water produced from these methods may be reinjected depending on the regulatory effluent requirements, the economic constraints and the available treatment options.

2 Literature Review

Membranes are slender layers of semi-permeable materials such as polymer or ceramic. Contaminants in feed water can be separated by application of pressure difference across the membrane [29]. These processes are mostly used for discarding microorganisms, particulates, bacteria and organic and inorganic substances, which transfuse colour, tastes and odours to water. Capital and operating expenses are diminishing due to emerging new technologies in module designing and membrane production [5]. The process of membrane separation through driving pressure is categorized into four types: Reverse osmosis, Nanofiltration, Ultra-Filtration and Micro-Filtration.

Reverse osmosis (RO) operates on principles of applying a pressure differential between a semi-permeable membrane that is larger than the osmotic pressure between the two fluids [3]. The semi-permeable membrane is considered nonporous and is generally a thin asymmetric material. Pre-treatment using other methods of filtration is often employed in oil and gas wastewater treatment to decrease the rate of fouling [24]. Nanofiltration (NF) membranes are selective membranes allowing ionic solutes smaller than 1 nm, as the name suggests. NF membranes can be used in place of RO when certain monovalent species are allowed in the effluent. With oil and gas applications, the limits for effluent species vary depending on the end use (reinjection, recycle for irrigation, or surface/aquifer discharge) [4, 33]. Ultrafiltration (UF) is completed by using porous membranes which separate macro-molecules, colloids and large dissolved solids from the product stream. The main method of separation is based on sieving contaminants based on size and surface charge of particles. Ultrafiltration membrane pores generally range from 1nm to 100nm. Lastly, microfiltration is used to separate impurities larger than 0.1–10 μm. Particles and biological contaminates including bacteria and viruses are among the species remaining in the retentate [22].

The efficient method for treatment of water with the residuals of oil is ultrafiltration. He [11] with studies showed reductions of 90% for chemical oxygen demand (COD), 98% for total organic carbon (TOC) and effluent concentrations of oil <1 mg/L [17]. And it's another advantage is that all the types of modules can be used in case of ultrafiltration. The material of the membrane also plays an important role in filtration of particles [9].

Membranes can be framed from polymeric and inorganic materials. The polymeric materials can be distinguished on the basis of certain properties like molecular weight, flexibility of chain, interaction of chain, which are like polytetrafluoroethylene, Teflon, polypropylene, polyethylene and cellulose esters [23, 25]. In case of high-temperature processes, it is efficient to use membranes made of ceramics, which are produced from materials such as nitrides, oxides, or carbides of metals such as zirconium (ZrO_2), alumina (Al_2O_3) and silicon carbide (SiC) [14, 32].

Different methods of enhancing performance of UF membranes include modification of materials. A study conducted on polyvinylidene fluoride composite membranes resulted in improved performance of flux and selectivity when modified with nano-sized, inorganic alumina [17]. While maintaining consistent transmembrane pressure at 0.1 MPa, the flux increased by nearly double the unmodified membrane flux. Quality of effluent based on COD, oil content, TOC, turbidity and suspended solids also improved with increased or maintained rejection rates of all categories. Furthermore, the study determined the modification had strengthened the antifouling performance of membrane due to the resulting hydrophobicity of the inorganic alumina. Both reversible and irreversible fouling was analysed for the modified membrane and showed improved permeate flux recovery from backwashing in comparison to the unmodified membrane. In using membrane washing fluids ranging from pure water to high pH solutions, the recovery ranged from 91.5 to 100%. Their study concluded that most fouling was due to colloid layer formation on the membrane and the remaining is irreversible by pure water backwashing.

While the study tested the membrane for quality of effluent, permeate flux and fouling, several questions remain to be answered [17]. One of which is the effect of temperature and pressure on modified membrane performance. Since the study was conducted at 30 °C, future tests on the modification should evaluate the range of temperatures and pressures on the material. Industrial application of UF may operate at high temperatures such as in SAGD recycled water treatments [1, 31]. Furthermore, the impact of varying contaminant concentrations was not discussed in the study. For example, high salinity may influence the rate and degree of irreversible fouling. Future research could evaluate the modification using nano-alumina on ceramic membranes which have more robustness against thermal and chemical degradation [1].

A pilot study on multichannel ceramic UF membranes (ZrO_2) was conducted using produced water from offshore oilfield produced water [30]. The purpose of the study is to determine the ability to use ceramic UF membranes to treat produced water within the limits required for water reinjection. The main contaminants of interest are suspended solids and hydrocarbon content which cannot be addressed by traditional methods such as hydrocyclones and dissolved gas floaters [6]. Permeate from the UF membrane had no detectable suspended solids and >3 mg of hydrocarbons, which is below the local limit of reinjection. Long-term evaluation of membranes showed resulting flux between 293 and 570 $L/h^{-1}m^{-2}bar^{-1}$, a value comparably higher than literature values determined with synthetic produced water [26]. The observation may be due to higher salinity of real produced water or membrane properties. Fouling on the membrane reacted more favourably with acid backwashing in comparison to alkaline or pure water which suggests fouling is due to inorganic compounds.

Additional research on the application of ceramic membranes for the produced water reinjection could be conducted on other sources and end use. Produced water from other production methods utilizing reinjection such as enhanced oil recovery, hydraulic fracturing [7] and SAGD for oil sands [18] have varying concentrations of contaminants and reinjection limits. Further research in these areas will determine the necessary pre-treatment required and specific backwashing techniques. At the stage of pilot testing, technological economic assessments must also be conducted to estimate the feasibility of wastewater treatment using ceramic membranes with necessary supplemental process units. An analysis on economics could evaluate the trade-offs between materials for membranes, extent of pre-treatment for UF, and scaled costs for capital and operating expenditures [15].

Some of the findings show that the total removal percentage of oil content was 93% when microfiltration was used as pre-treatment and 99.5% removal when ultrafiltration and Nanofiltration were used coincidently in the final staged treatment [2].

Even though all these results are based on a wide range of variables, a key comparison would have been to include the flow patterns to properly decipher the most effective and efficient membrane to use in oil and gas industry.

From the literature review, it is clearly evident that Ultrafiltration is the best methodology for the treatment of produced water. In this type of membrane filtration, forces like concentration gradient and pressure yield in separation of impurities ranging from 10^3 to 10^6 Da in size through a semi-permeable membrane. When compared to prevalent distillation methods, it is the most productive method for impure water treatment, mainly for oilfield-produced water due to its high efficiency of oil removal, low capital and operating cost, and constant quality of water with respect to the removal of oil and particles.

This provides a potential of eliminating high molecular weight substances, oil molecules, colloidal particles and all the microbiological species.

3 Methodology

Produced water samples are collected from KG basin. Different types of membranes as shown in Table 1 are tested to determine the ideal membrane for treatment of produced water. Pressure is varied as shown in Table 2 to determine the optimum operating pressure of membrane.

The flow sheet (Fig. 1) describes the experimental setup.

Procedure:

- Feed water is heated to the required temperature using a heater in a feed tank. The temperature is monitored using a temperature indicator.
- The water in a feed tank is mixed by a mixer with a rpm of 100.
- Feed water is pumped to the membrane module through a pump.
- The pressure of the membrane module is visualized by a pressure indicator.
- Pressure and temperature are varied for different samples.
- Temperature is varied from 30 to 40 °C, PH is 5–9, and pressure is varied according to the values in Table 2.
- Flow rates are controlled by using the valves.
- Permeate is collected in the permeate collection tank and sent for further analysis.
- Backwash water and retentate are collected in the retentate collection tank.

Table 1 Characteristics of membranes used

S. no.	Membrane material	pH range	Max. pressure (bar)
1	Polyvinylidene difluoride	1.6–12.2	11
2	Polyethersulphone	1.6–12.2	11
3	Polyamide	1.4–9.6	11
4	Zirconium oxide	2.2–13.2	11

Table 2 Values obtained at given temperature (35 °C) and pressure

Membrane	Parameters (units)						
	Pressure (Bar)	Turbidity (NTU)	Oil & grease (PPM)	TSS	TDS	COD	TOC
Polyvinylidene difluoride	5	0.6	0.15	19	20,395	17	22
	7	0.9	0.2	22	21,546	21	26
	9	0.7	0.18	21	21,121	19	23
	11	0.4	0.12	17	19,896	16	19
Polyethersulphone	5	0.7	0.17	19	20,389	18	23
	7	0.5	0.15	17	19,698	15	20
	9	0.6	0.17	19	20,389	18	23
	11	0.4	0.1	16	18,985	13	17
Polyamide	5	<0.3	<0.1	14	16,791	11	15
	7	<0.3	<0.1	14	16,791	11	15
	9	<0.3	<0.1	13	14,897	10	14
	11	<0.3	<0.1	11	12,258	9	13
Zirconium oxide	5	<0.3	<0.1	11	12,139	9	14
	7	<0.3	<0.1	10	11,437	8	11
	9	<0.3	<0.1	12	13,578	10	13
	11	<0.3	<0.1	11	12,139	9	11

Note Abbreviations used °C—degree Celcius; PPM—Parts Per Million; NTU—Nephelometric Turbidity unit

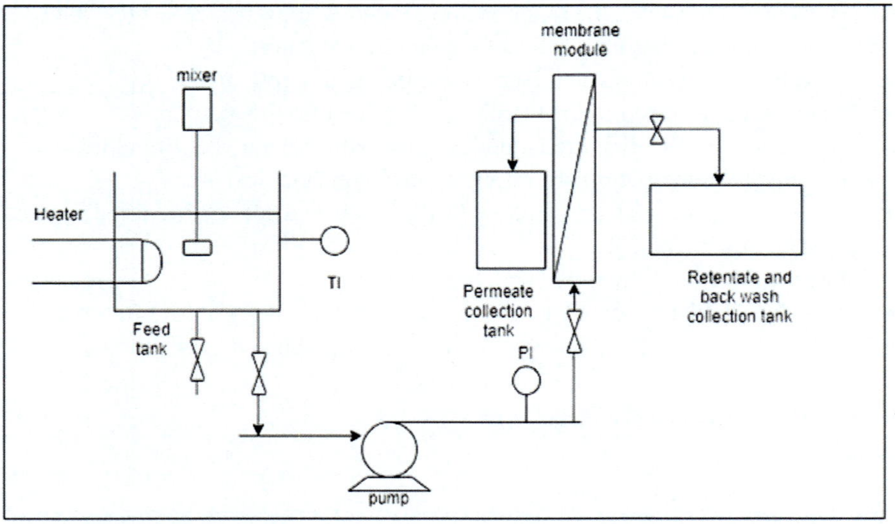

Fig. 1 Flowsheet of experimental setup

4 Experimental Data Analysis of Permeate

Permeate on the given Temperature (Temp) and Pressure is tested for Total Suspended Solids (TSS), Total Dissolved Solids (TDS), Total Organic Carbon (TOC), Chemical Oxygen Demand (COD), Turbidity by standard testing procedures in the lab and the values are tabulated in (Table 2).

5 Results and Discussions

Figures 2, 3, 4 and 5 show the conductivity of permeate as the pressure is varied. In Fig. 2, the conductivity of permeate increased at 7 bar due to the accumulation of particles on membrane. Recovery rate increases with the pressure. But when membrane is operated at high pressure for a prolonged period the membrane gets damaged. So, the ultrafiltration is carried at optimum pressure. In order to avoid pore blocking the membrane is back washed and the water is collected in the backwash water collection tank. Figures 2, 3, 4 and 5 show the conductivity of ultrapure water by using different membranes. X-axis represents pressure applied, and Y-axis represents the conductivity of permeate. In Fig. 3, the conductivity is increased at a pressure of 9 bar. It indicates that the membrane is blocked at that pressure. In Fig. 4, the graph follows the linear trend. It shows that there is no problem of pore blockage in polyamide membrane. In Fig. 5, the conductivity increased while operating at 9 bars. And at 11 bar the conductivity is reduced. But if the membrane is operated at 11 bar the shelf life of membrane decreases significantly. Moreover, the efficiency at 11 bar is less than that of 7 bar, so 7 bar is the optimum pressure for the operation of zirconium oxide. The efficiency of treatment is shown in Table 3.

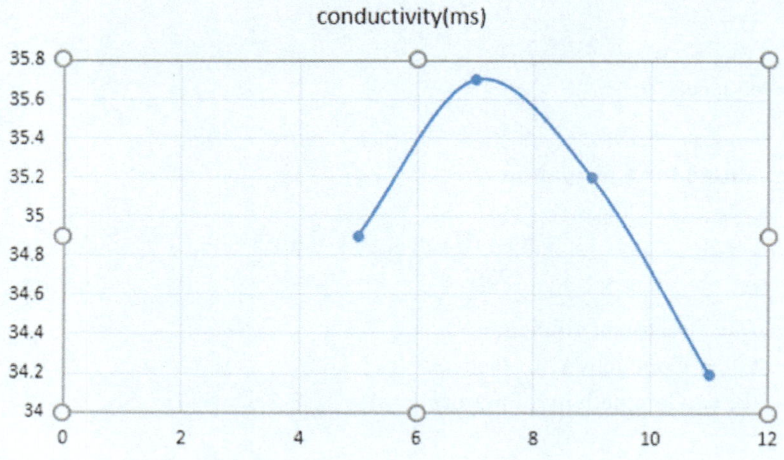

Fig. 2 Polyvinylidene difluoride

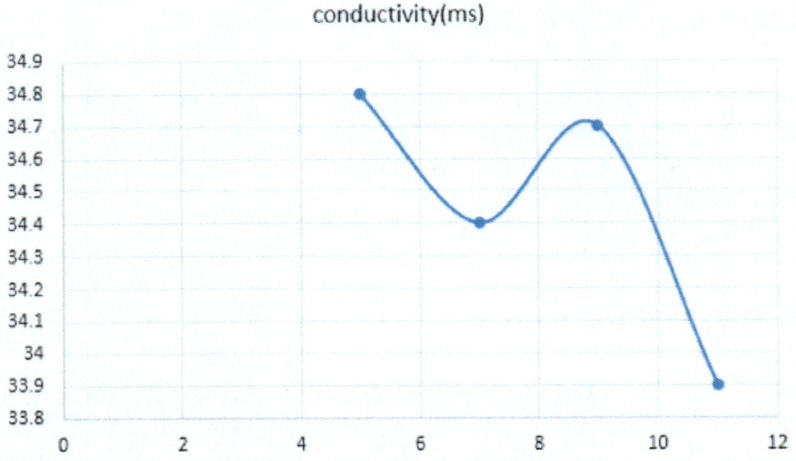

Fig. 3 Polyether sulphone

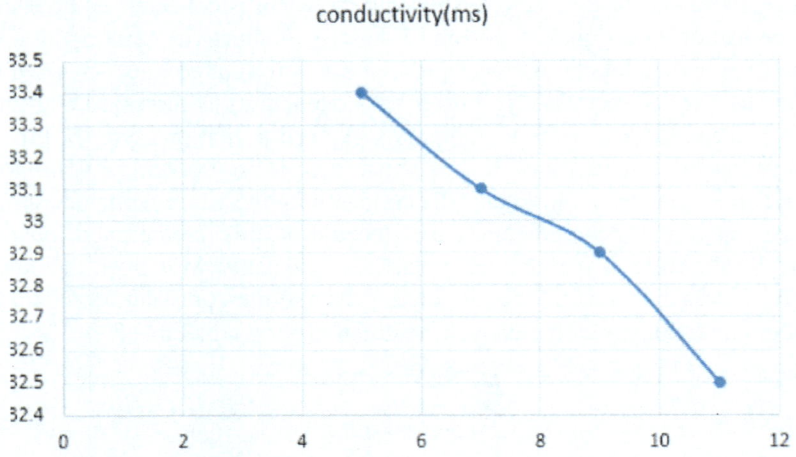

Fig. 4 Polyamide

Formula used for Calculation:

$$R = [(C_i - C_0)/C_i] * 100$$

where, $R =$ treatment efficiency
$C_i =$ Conductivity of feed
$C_0 =$ Conductivity of permeate

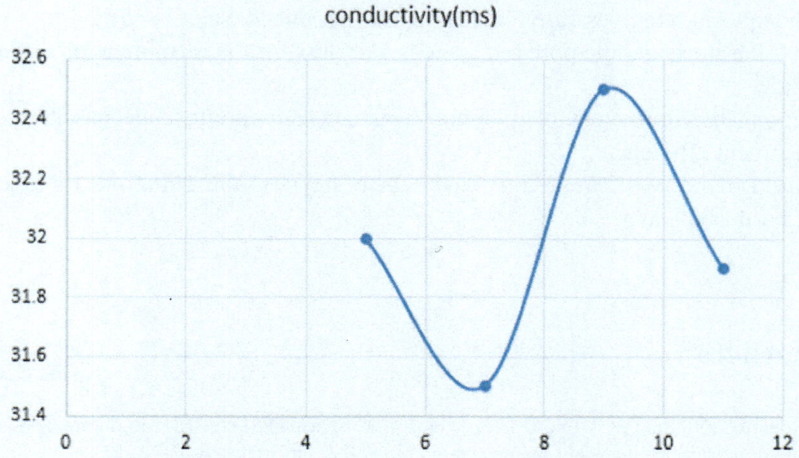

Fig. 5 Zirconium oxide

Table 3 Efficiency of the treatment

S. no.	Membrane	Pressure (bar)	Treatment efficiency (%)
1	Polyvinylidene diflouride	5	32.8
		7	31.3
		9	32.2
		11	34.2
2	Polyethersulphone	5	33.0
		7	33.8
		9	33.2
		11	34.8
3	Polyamide	5	35.7
		7	36.3
		9	36.7
		11	37.5
4	Zirconium oxide	5	38.4
		7	39.4
		9	37.5
		11	38.6

6 Conclusion

The study concludes the following important points:

- Polyvinylidene difluoride, polyethersulphone, polyamide and zirconium oxide membranes are tested at varied pressures.

- The optimum temperature for treatment of produced water is 35 °C.
- At 7 bar pressure, the pore blockage is very less for the membrane of Zirconium oxide.
- Among the four types of membranes tested, Zirconium oxide has the highest treatment efficiency.
- Based on the tests, Zirconium oxide is the best type of the membrane for treatment of produced water.

References

1. Adham S, Hussain A, Minier-Matar J, Janson A, Sharma RJD (2018) Membrane applications and opportunities for water management in the oil & gas industry, 440:2–17
2. Avnimelech Y, Ritvo G, Meijer LE, Kochba M (2001) Water content, organic carbon and dry bulk density in flooded sediments, 25(1):25–33
3. Chen JP, Wang v, Yang L, Zheng Y-M (2011) Desalination of seawater by thermal distillation and electrodialysis technologies. Membrane and desalination technologies. Springer, pp 525–558
4. Conrad CL, Yin YB, Hanna T, Atkinson AJ, Alvarez PJ, Tekavec TN, Reynolds MA, Wong MSJWR (2020) Fit-for-purpose treatment goals for produced waters in shale oil and gas fields, 173:115467
5. Drioli E, Stankiewicz AI, Macedonio F (2011) Membrane engineering in process intensification—an overview, 380(1–2): 1–8
6. Ebrahimi M, Willershausen D, Ashaghi KS, Engel L, Placido L, Mund P, Bolduan P, Czermak PJD (2010) Investigations on the use of different ceramic membranes for efficient oil-field produced water treatment, 250(3):991–996
7. Estrada JM, Bhamidimarri RJF (2016) A review of the issues and treatment options for wastewater from shale gas extraction by hydraulic fracturing, 182:292–303
8. Fakhru'l-Razi A, Pendashteh A, Abdullah LC, Biak DRA, Madaeni SS, Abidin ZZ (2009) Review of technologies for oil and gas produced water treatment, 170(2–3):530–551
9. Filteration S (2019) Module configurations & processes
10. Gray M, Deutsch N, Marrero M (2017) Current regulatory issues regarding hydraulic fracturing. In: SPE hydraulic fracturing technology conference and exhibition, OnePetro
11. He Y, Jiang Z-W (2008) Technology review: treating oilfield wastewater, 45(5):14–16
12. Healy RW, Alley WM, Engle MA, McMahon PB, Bales JD, U.S. Geological Survey Circular (2015) The water-energy nexus: an earth science perspective, 1407
13. Intelligence, G. W., I. Yearbook, G. W. Summit and R. J. G. W. I. Card (2011). Global water intelligence, 12(10):1–72
14. Kayvani Fard A, McKay G, Buekenhoudt A, Al Sulaiti H, Motmans F, Khraisheh M, Atieh M (2018) Inorganic membranes: preparation and application for water treatment and desalination, 11(1):74
15. Kehrein P, Jafari M, Slagt M, Cornelissen E, Osseweijer P, Posada J, van Loosdrecht M (2021) A techno-economic analysis of membrane-based advanced treatment processes for the reuse of municipal wastewater, 11(4):705–725
16. Leos JZ, Zydney AL (2017) Microfiltration and ultrafiltration: principles and applications. Routledge
17. Li YS, Yan L, Xiang CB, Hong LJ (2006) Treatment of oily wastewater by organic–inorganic composite tubular ultrafiltration (UF) membranes, 196(1–3):76–83
18. Mahaffey A, Dubé M (2017) Review of the composition and toxicity of oil sands process-affected water, 25(1):97–114

19. Mulder M, Mulder J (1996). Basic principles of membrane technology. Springer Science & Business Media
20. Nandan A, Yadav BP, Baksi S, Bose D (2017) Recent scenario of solid waste management in India, (66):56–74
21. NETL (2019) Crosscutting research program
22. Ozofor IH (2020) Towards efficient water treatment: mechanism of colloidal fouling of ultrafiltration membranes. Michigan Technological University
23. Salahi A, Gheshlaghi A, Mohammadi T, Madaeni SS (2010) Experimental performance evaluation of polymeric membranes for treatment of an industrial oily wastewater, 262(1–3):235–242
24. Shams Ashaghi K, Ebrahimi M, Czermak P (2007) Ceramic ultra-and nanofiltration membranes for oilfield produced water treatment: a mini review, 1(1)
25. Strathmann H (1981) Membrane separation processes, 9(1–2):121–189
26. Tansakul C (2009) Procédés hybrides à membranes pour le prétraitement d'eau de mer avant dessalement par osmose inverse. Institut National des Sciences Appliquées de Toulouse
27. USGS (2019) Energy resources program
28. Veil JA, Puder MG, Elcock D (2004) A white paper describing produced water from production of crude oil, natural gas, and coal bed methane. Argonne National Lab., IL (US)
29. Wenten I (2005) Recent development in membrane and its industrial applications. Membrane Technology in Oil and Gas Industry
30. Weschenfelder SE, Louvisse AM, Borges CP, Meabe E, Izquierdo J, Campos JC (2015) Evaluation of ceramic membranes for oilfield produced water treatment aiming reinjection in offshore units, 131:51–57
31. Ganguly M, Aynyas R, Nandan A, Mondal P (2018) Hazardous area map: an approach of sustainable urban planning and industrial development—a review. Nat Haza 91(3):1385–1405
32. Jha K, Nandan A, Siddiqui NA, Mondal P (2020) Sources of heavy metal in indoor air quality. In: Advances in Air Pollution Profiling and Control (pp 203–210). Springer, Singapore
33. Mondal P, Nandan A, Siddiqui NA, Yadav BP (2014) Impact of Soak Pit on Groundwater Table. Environ Pollut Control J 18:12–17
34. Nandan A, Yadav BP, Baksi S, Bose D (2017) Assessment of water footprint in paper & pulp industry & its impact on sustainability. World Sci News 64:84–98
35. Ora A, Nandan A, Kumar A (2017) Hazard identification of chemical mixing plant through hazop study. Int J Advan Res Develop 2(3)
36. Padgelwar S, Nandan A, Mishra AK (2021) Plastic waste management and current scenario in India: a review. Int J Environ Analy Chem 101(13):1894–1906
37. Rana D, Joshi N, (2023) Assessment of Human Intervention on Ecology of River—A Case Study. In: International Conference on Advances and Innovations in Recycling Engineering (pp 51–70). Springer, Singapore
38. Subuddhi SP, Kansal A, Pandey P, Rana D, Ghoshal, T Study of Impact of Influx Waste Water Generated at Vehicle Service Centre on River Water Quality Characteristics. Damini and Ghoshal, Tarumay, Study of Impact of Influx Waste Water Generated at Vehicle Service Centre on River Water Quality Characteristics

Investigation of Organic Carbon Contamination in Himalayan Rivers of Uttarakhand

Damini Rana, Neha Tiwari, Sudarshan S. Pal, S. P. Subudhi, Ankur Kansal, and Namita Joshi

Abstract The Himalayan state of Uttarakhand is vulnerable to ongoing changes which are either climate changes or continuous migration activities. This is a newly formed state and somehow unstable with respect to developmental projects. Rivers originating from Himalayas are facing the issues of deteriorating water quality day by day. River Ganga originating from Uttarakhand holds a National religious importance as well as being the lifeline to various organisms. But, according to previous studies, river quality is noticed to be affected due to natural and anthropogenic activities. This paper focuses on the pollution by organic carbon, in the Uttarakhand region of river Ganga and its tributaries. Therefore, the work aims to measure the Total Organic Carbon (TOC) in the respective river using TOC Analyser. Four different locations such as Gangotri, Nandprayag, Srinagar and Rishikesh are considered as the river water sampling sites. High concentration of carbon was observed in Nandprayag region though it has very less population in comparison to Rishikesh. Conclusion of the study shows, though anthropogenic activities affect the quality of river water but the total flow of river is equally important to maintain the healthy ecological activities in it. Also, the values of TOC in the river at the Gangotri region with very less population were found similar to the values of distilled water which justifies the fact that anthropogenic activities are majorly responsible for the higher content of organic contamination in the river water.

D. Rana (✉)
Department of Environmental Sciences, KGC, Gurukula Kangri (Deemed to Be University), Haridwar, India
e-mail: daminiranaofficial@gmail.com

Uttarakhand and Central Laboratory, Uttarakhand Pollution Control Board, Dehradun, India

N. Tiwari · N. Joshi
Department of Environmental Sciences, Gurukula Kangri (Deemed to Be University), Kanya Gurukula Campus, Haridwar, Uttarakhand, India

S. S. Pal · A. Kansal
Uttarakhand Pollution Control Board, Dehradun, India

S. P. Subudhi
Dehradun and Uttarakhand Pollution Control Board, DIT University, Dehradun, India

N. A. Siddiqui et al. (eds.), *Advances in Waste Management*, Lecture Notes in Civil Engineering 301, https://doi.org/10.1007/978-981-19-7506-6_8

Keywords Organic carbon · River Ganga · Anthropogenic activities · Water pollution · River ecology

1 Introduction

Uttarakhand, the newly formed state, situated in the Himalayas is vulnerable to the changes which are either climate changes or the continuous migration activities. The state is somehow unstable with respect to increasing developmental activities. Rivers originating from Himalayas are facing issues of deteriorating water quality day by day. River Ganga is the largest river in India. It serves as the lifeline for a large population in India as it covers approx. 26% area of the country. The river Ganga has a total length of 2,525 km rising from the Himalayas in Uttarakhand. As per [37, 27] the entire Ganga is made of many tributaries and one of the major tributary originating from Gaumukh-Gangotri glacier is called Bhagirathi which joins Alaknanda at Devprayag forming the Ganga River. The other largest tributary of Ganga is river Yamuna which after joining at Allahabad, finally makes its way towards the Bay of Bengal [31, 32].

Ganga serves as home for around 350 species of fish in which several are endemic [20]. The river also serves as the habitat for many endangered aquatic animals like gharial crocodile, mugger crocodile, various turtles and famed faunal member, the Ganges River Dolphin. These species are seriously threatened due to pollution and rise in organic matter and toxic substances [1]. Around 40% of the population of the country lives in the Ganga basin. The Ganges basin is the home for 400 million people with a population density of 1,000 inhabitants per sq. miles, which makes it the most heavily populated river basin in the world and the main source of pollution is due to the discharge of untreated wastewater from fast-growing human habitats along the river banks [18, 39].

According to an earlier research, the Himalayan River system contributes between 10 and 20% of the total worldwide flux of contemporary organic carbon to the oceans [25]. The Total organic carbon and inorganic carbon are together called Total carbon. The presence of the carbon pollutants in the river water would degrade the system entirely and make it unfit for drinking after disinfection or even for bathing purposes. The organic pollutants mainly contain sewages while inorganic contains the mixture of minerals, salts as well as heavy metals etc, [7]. The sources of organic content (carbon) can be natural things, industrial waste and effluents, leachates from solid waste disposal sites, rural and urban runoff and atmospheric fall out [4, 22].

The organic pollution occurs due the presence of organic compounds in the river bed in large quantities. Such compounds act like a substrate for the microbes in the water [8]. So, during the decomposition process the Dissolved oxygen (DO) will be consumed in an extent that it cannot be replenished easily, which leads to oxygen depletion in the aquatic environment, affecting the stream adversely [16]. The organic effluent would also lead to adverse effects. Large quantities of suspended particles in the river water will reduce the light availability, which in turn affect the photosynthesis and alter the river bed characteristics and will make the habitat unsuitable for the dependents [14].

Main source of organic carbon is clearly mentioned by Bedding et al. [6, 13, 19, 24, 41] in the river water as:

(i) Natural substances, are the things associated with the life cycle of living organisms. It could be the decayed plants inside the river or released to the river through run-off or leaching, animal excreta, human wastes, food wastes or any other metabolic byproducts.
(ii) Industrial waste and effluents produce large amount of organic carbon contents in river water. It mainly includes aliphatic and aromatic hydrocarbons, petroleum hydrocarbons, benzenes, PAH, organ chlorine pesticides, PCBs and esters.
(iii) Leachates from solid waste disposal sites, comes into the river water mainly by two considerations. One is through surface run-off and the other is by the penetration of the leachates to the ground water table.
(iv) Through urban and rural run-off, compounds like hydrocarbons, PAH, fatty acids, ketones, ester plasticizers and other compounds, would reach the river.
(v) Atmospheric fall-out is another source of pollution that may reach into the water system.

Total Organic Carbon is relatively a newly introduced parameter (i.e., 1970), which is based on carbon and provides clear quantification of the organic carbon (Organic contamination) present in the water. The amount of carbon in a freshwater stream indicates the level of organic character of the stream [3]. As the organic content increases more oxygen on the water body is consumed as the microorganisms increase due to the presence of this organic content [26].

Previous studies suggest that Ganga water pollution is mainly due to industrial wastes and domestic waste [10, 12, 29, 34]. Ind.ustries like paper mills, fertilizer plants, and chemical plants are also responsible for water pollution at river Ganga [2, 29, 33, 37, 23]. Around 2500 MLD of wastewater dumped on Ganga are from industries which is 20% of the total waste water dumped in to the river. The domestic waste includes mostly organic waste, sewage, food waste human and animal waste and remains [36]. The sacred practice of depositing human burial remains in the Ganges also results in health issues due to the dumping of partially cremated cadavers [31, 32, 35].

The river system plays a crucial role in the worldwide biogeochemical cycle of carbon through the transport and destiny of organic carbon [28, 40]. Thus, the study aims to measure the total organic carbon (TOC) in river Ganga and its tributaries. Analysis of TOC is essential for determining the water quality of river streamflow as it can be regarded as one of the significant indicators [15] for healthy ecological conditions.

2 Materials and Methodology

The main aim of the project is to measure the TOC present in the river Ganga water in Uttarakhand and to investigate the sources of the organic pollution. So, the project needs to be scheduled in a series of steps to achieve the right results as represented in the Flowchart below (Fig. 1).

1. **Selecting sampling sites and sampling frequencies**: Four sampling sites chosen for the study are: Gangotri, Nandprayag, Srinagar and Rishikesh shown in Fig. 2. In a pilot study Gaumukh-Gangotri was found very less populated and less affected by anthropogenic activities in most of the months in comparison to the other three sites. Thus, Gangotri was considered to be the benchmark for the other three sampling sites.

Sample were collected from the respective sites on a monthly basis in the year 2019. One major factor under consideration was population because population is one of the significant reasons behind the sewage pollution dumped into the river. The second factor was about the locations where more amount of sewage is getting deposited. While the third factor to be considered was the sample collections from different tributaries to get the idea of total organic carbon content individually.

2. **Sampling procedure**: The sampling procedure is based on the literature reviewed [17]. According to this it is mentioned as collecting water from the river by a can directly is secondary, as in primary it is to collect the water from the center of

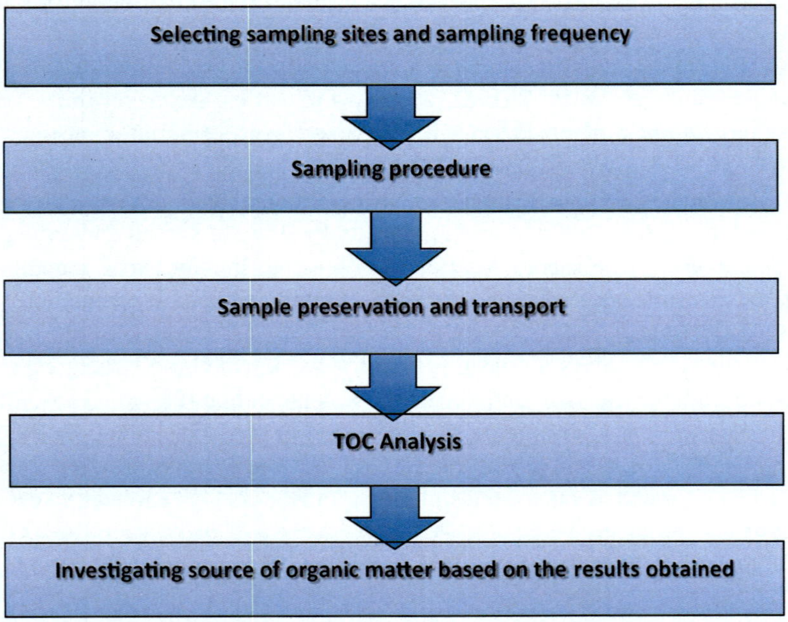

Fig. 1 Flowchart representing steps of work methodology

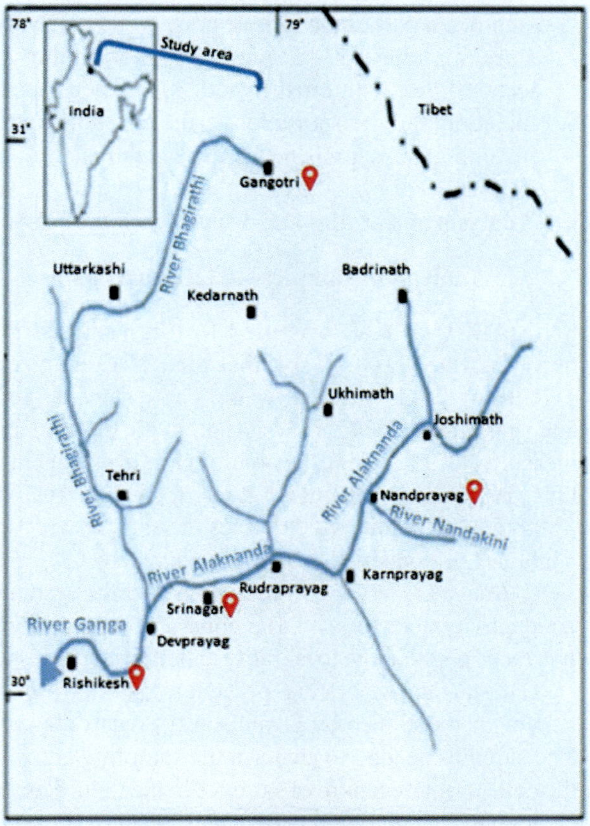

Fig. 2 Sampling sites of the study conducted

the river, if a bridge accessibility is there. Therefore, with a rope and bucket as accessories, by standing on the center of the bridge, the sampling team can drop the bucket to the water river by extending the rope—sampling of Nandprayag and Rishikesh done in this way.

But due to circumstances in many places the usage of the secondary method comes into picture, whereas such bridge facilities won't be available in many locations. In such locations like Gangotri and Srinagar, by considering personal safety in the first place, the sampling person went in river till knee level water and immersed the sampling bottle approximately to a depth of 15 cm from the water surface, as the top layer of 0–5 cm from the surface water can have impurities, oil contaminations etc. The bottle was capped carefully to avoid any human contact with the collected sample or even the tip of the bottle. After successfully completing the water sample collection from the river, the sampling team labeled the date, time and name of site on the container with blue ball pen and white sticking paper.

3. **Sample preservation and transport**: When the time gap between the sampling and analysis increases, it degrades the sample leading to poor results. The most

significant part of the sample preservation is to keep the sample at a cool temperature in a range of +2 and +5 by storing the sample in a refrigerator or in an icebox filled with crushed ice. Also, keep the sample away from the daylight including the transportation to the lab to be kept in a chilled condition. During travelling, the sample needs to be kept inside a thermal flask partially filled with crushed ice.

4. **Analysis of samples and Identification of objectives:**

   A. **Analysis of samples—TOC analysis**

SHIMADZU TOC-L was used for the analysis of TOC. The Operation Manual of SHIMADZU TOC-L specifies that, TOC analyzer is the machine to measure the TOC/IC of the Ganga River water. The system has automatic sampling, acidification, and sparging for TOC analysis, automatic dilution and auto calibration facilities. To measure the TOC value, the organic carbon present in the river water must be oxidized thus, TC needs to be converted into CO_2. In SHIMADZU, it uses 680° Combustion Catalytic Oxidation, which mixes the sample with an oxidation catalyst in a heated container to convert the organic content to CO_2. Finally, in the detection system TOC Analyzers use two types of detection systems, non-dispersive infrared and conductivity detectors. NDIRs consist of a light source, cell and a detection portion. It is more popular due to stability and interference issues with conductivity detectors.

The standard operating procedure for performing water sampling in river or streams is mentioned by Danielson [9] regarding the sampling period and location. The supplies needed to perform the sampling are water quality kits, gloves etc. and the collection procedure of samples in the field. The TOC analysis of the samples can be done by calculating the Total Carbon (TC) and subtracting the value of Inorganic Carbon (IC) and this method is referred as the TC-IC method. For this analysis a TOC analyser should be used. If it is a PC controlled one, the TOC-L sample editor software as an interface to make the TOC analyser work is used. The TOC analyzer works on three phases sampling, acidification, oxidation, detection and quantification. The acidification involves addition of acid and flushing of inert gas to remove all other dissolved gas. While the second stage involves addition of oxygen in the sample to convert the rest of the carbon in the sample to carbon dioxide. The oxidation can be done using six methods and the most common and accurate method is the High temperature catalytic oxidation (HTCO) method. And finally for the detection and quantification of the desired result, the conductivity method and NDIR method are used.

B. **Identification of the following objectives:**

   (1) Investigating the relation between population size and organic pollution
   (2) Investigating the relation of sewage water deposited into river and organic pollution
   (3) Investigating the sources of higher organic content in the river Ganga.

3 Results and Discussion

A. TOC Analysis—Results

Samples were analyzed for four locations (Gangotri, Nandprayag, Srinagar and Rishikesh), Gangotri being the benchmark. The yearly average value of the total organic carbon analyzed on a monthly basis is calculated to get the clear picture of organic contamination. Also, before the analysis of the samples it is mandatory to perform the analysis with distilled water. Thus, distilled water is also tested as a sample reference against polluted river water which gave the result of the TOC value as 0.543 mg/L. This paper verifies, how much TOC is higher or lower in all the 3 samples as compared to the upstream Himalayan river. The measured values of TOC from various locations are mentioned in Table 1 and graphically represented in Fig. 3.

Therefore, it is concluded from the analysis that in Nandprayag, river water has the highest TOC value, while Gangotri water has the lowest TOC value (0.445 mg/L) which is even lesser than the TOC value of distilled water (0.543 mg/L).

B. Investigation based on the results obtained and sampling site observation.

One of major target of this project analysis of TOC was to find the sources of organic matter and to find the relation between different parameters and organic matter on

Table 1 Measured values of TOC at various location

S. no.	Sample collected from the source	TOC values obtained (mg/L)
1	Distilled water (as reference)	0.543
2	Gangotri	0.445
3	Nandprayag	17.400
4	Srinagar	2.130
5	Rishikesh	1.973

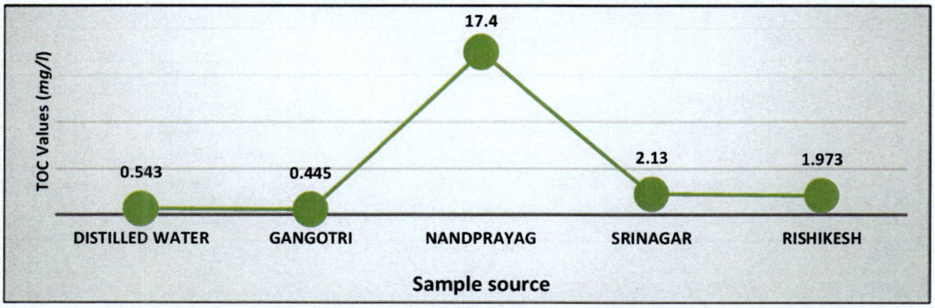

Fig. 3 Graphical representation of TOC values obtained from different sample sources

the river. The parameters include population size, sewage water deposition (flow rate of sewage water) on the river, industries that depend and other geographic factors.

The TOC value of Nandprayag is too higher than expected, it could be due to the Gopeswar town situated on the uphill side of Nandprayag, as the activities in Chamoli-Gopeshwar towns can also affect the Nandakini tributary. So, it will be more appropriate to add the parameters of Chamoli-Gopeshwar with Nandprayag to get more accurate comparison.

(1) **Relation between Population Size and Organic Matter Content**

The population data residing in the nearby river of each sampling location is tabulated below. In the following Table 2, the population of Nandprayag (which is 1433) and population of Chamoli-Gopeshwar (which is 21,447) is added up. Figure 4 is the graphical representation of variation in TOC values with respect to the population size of the sampling region.

From this graph we could not get a linear relation with population and organic pollution in the river. As the value of TOC in Nandprayag is higher than the TOC value of Rishikesh despite having a much lesser population. It is a contradicting factor, but these unexpected results may be due to the amount of water flowing through the respective sites. To be clear, the width, depth, flow rate of the Ganga

Table 2 Population residing nearby selected sampling locations

S. no.	Sampling site	Population
1	Gangotri	606
2	Nandprayag (Chamoli-Gopeshwar tows included)	22,880
3	Srinagar	20,115
4	Rishikesh	102,138

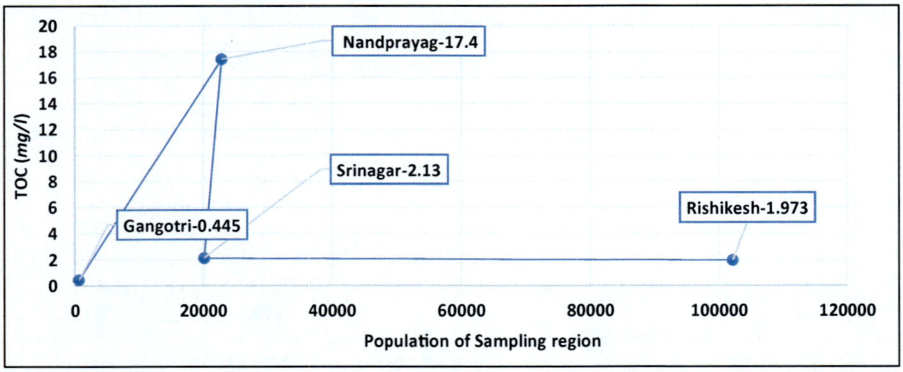

Fig. 4 Curve based on population and TOC values

River in the Rishikesh region is in peak as compared to the same parameters of the Nandprayag.

Therefore, even though the pollution and sewage factors are quite higher, due to the above-mentioned parameters of river, it is not getting reflected. Moreover, any unidentified geographic factors can also show such kind of variations in the sites. i.e., minerals or any other deposits in the rocks can even increase the TOC value [5] in Nandprayag region.

(2) Relation between Organic Matter Content and Sewage Water Deposition (Flow Rate of Sewage Water)

The sites chosen are significant points, because several sewage nallas are located in these spots that dump waste water into Ganga River. Hence, these flow rates of sewages are being plotted against the TOC values obtained so as to find any relation between the amount of sewage water deposited on the river and its TOC values. The sites in which drains are present and its flow rate is tabulated below in Table 3 with graphical representation in Fig. 5.

From this graph we couldn't get a linear relation with sewage water deposition and organic pollution in the river. As the value of TOC in Nandprayag is higher than TOC value of Rishikesh despite having much lesser sewage water deposition. In

Table 3 Number of drains and their average flow rate for respective sampling sites

S. no.	Sampling site	No of drains	Average flow rate in MLD
1	Gangotri	0	0
2	Nandprayag (Chamoli-Gopeshwar towns included)	1 + 4	14.01MLD
3	Srinagar	3	5.93MLD (2.56 + 3.02 + 0.35)
4	Rishikesh	3	153.25MLD (8.97 + 60.05 + 84.23)

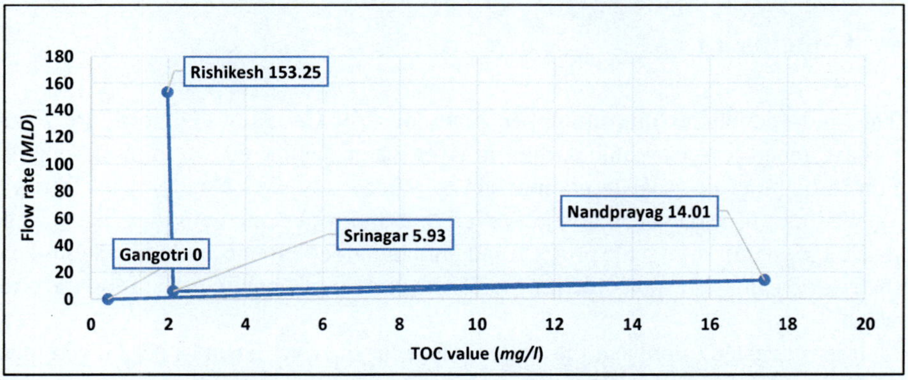

Fig. 5 Graph between sewage water deposition Flow rate and TOC values

previous studies the amount of organic content decreases with the increase in water flow of the river [11, 30]. Thus, large amount of water flowing with sufficient depth and width of the river in Rishikesh as compared to the river in Nandprayag can be the reason for results showing low TOC value in Rishikesh despite having much sewage deposition than the other sampling sites.

(3) **Identifying the sources of higher organic content in the river Ganga**

From the analysis, TOC of the origin of the Himalayan river at Gangotri showed a purity equal to distilled water and Nandprayag has unexpectedly shown the highest value of TOC among all the sampling locations. Absence of any polluting sources and very fewer human activities in Gangotri in most of the months can be the reason behind the less TOC content in the analysis. While Nandprayag and Gopeswar town (situated on the uphill side of Nandprayag) could be jointly responsible for the affecting the Nandakini tributary and also the unstudied geographic factors such as minerals or any other deposits in the respective area can also cause the same. While, Rishikesh having large number of domestic sewage-nallas, hotels and religious activities in compassion to other sampling locations has very less total organic carbon, which can be due to the large size of the river and flow characteristics of the river in Rishikesh region.

Though the value for TOC content and population was much higher in Nandprayag, Srinagar and Rishikesh than Gangotri it paved the way to the conclusion that anthropogenic activities in the heavily populated riverside locations are the main reason for the river pollution. But Rishikesh being heavily polluted among all sampling points and showing less organic content than Nandprayag justified that the anthropogenic activities alone are not the responsible factors for the presence of higher content of TOC in the sample but the sufficient flow of the river also holds a significant place in balancing the total organic carbon discharged from various sources.

4 Conclusion

The study concludes that the upper zones of river Ganga at Gangotri, are found to have much lesser organic carbon in river water quality due to minimal population throughout the year in comparison to other study sites Nandprayag, Srinagar and Rishikesh. It has been also observed that anthropogenic activities like sewage discharge, small-large scale projects and unplanned settlements including slums near the river basin are responsible for increased content of organic carbon ion the river water.

It is suggested to prevent the settlements near the river basin as per Government of India norms in order to maintain the river quality characteristics. Disposal of sewage and other waste materials into the river basin must be strictly prohibited by the authorities. Installation of Water Treatment Plants and vegetation nearby the river basin can be enhanced to maintain the river water ecology.

References

1. Aggarwal D, Kumar N, Dutta VJES (2020) Impact on endangered Gangetic dolphins due to construction of waterways on the river Ganga, India: an overview, 3(2):123–138
2. Arif Z, Sethy NK, Mishra PK, Verma B, Water Management: Resources and Scarcity (2021) Grossly polluting industries and their effect on water resources in India. 47–65
3. Asmala E, Haraguchi L, Markager S, Massicotte P, Riemann B, Staehr PA, Carstensen J (2018) Eutrophication leads to accumulation of recalcitrant autochthonous organic matter in coastal environment, 32(11):1673–1687
4. Babayemi JO, Ogundiran MB, Osibanjo O (2016). Overview of environmental hazards and health effects of pollution in developing countries: a case study of Nigeria, 26(1):51–71
5. Banerji US, Bhushan R, Joshi KB, Shaji J, Timothy Jull AJ (2021) Hydroclimate variability during the last two millennia from the mudflats of Diu Island, Western India, 56(7): 3584–3604
6. Bedding ND, McIntyre A, Perry R, Lester JN (1982) Organic contaminants in the aquatic environment I. Sources and occurrence, 25(2):143–167
7. Bolan N, Duraisamy VP (2003) Role of inorganic and organic soil amendments on immobilisation and phytoavailability of heavy metals: a review involving specific case studies, 41(3):533–555
8. Chan SS., Khoo KS, Chew KW, Ling TC, Show PL (2022) Recent advances biodegradation and biosorption of organic compounds from wastewater: Microalgae-bacteria consortium—a review, 344:126159
9. Danielson T (2014) Protrocols for collecting water grab samples in rivers, streams and freshwater wetlands
10. Dimri D, Daverey A, Kumar A, Sharma A, Monitoring and Management (2021) Monitoring water quality of River Ganga using multivariate techniques and WQI (Water Quality Index) in Western Himalayan region of Uttarakhand, India, 15:100375
11. Dutta S, Dwivedi A, Suresh Kumar M, Assessment (2018) Use of water quality index and multivariate statistical techniques for the assessment of spatial variations in water quality of a small river, **190**(12):1–17
12. Dutta V, Dubey D,. Kumar S (2020) Cleaning the River Ganga: Impact of lockdown on water quality and future implications on river rejuvenation strategies, 743:140756
13. Ganguly M, Aynyas R, Nandan A, Mondal P (2018) Hazardous area map: an approach of sustainable urban planning and industrial development—a review. Nat Hazards 91(3):1385–1405
14. Jones J, Duerdoth C, Collins A, Naden PS, Sear DA (2014) Interactions between diatoms and fine sediment, 28(3):1226–1237
15. Kim S, Maleki N, Rezaie-Balf M, Singh VP, Alizamir M, Kim NW, Lee J-T, Kisi O (2021) Assessment of the total organic carbon employing the different nature-inspired approaches in the Nakdong River, South Korea. Environ Monit Assess 193(7):445
16. Lemaire GG, Carnohan SA, Grand S, Mazel V, Bjerg PL, McKnight US (2021) Data-driven system dynamics model for simulating water quantity and quality in peri-urban streams, 13(21):3002
17. Madrid Y, Zayas Z, Danielson T, Ye, J. Beach, J. Martin, Senthilselvan AJIJO, Health P (2014) Protocols for collecting water grab samples in rivers, streams, and freshwater wetlands, 26(4):6442–6471
18. Meena S (2019) A case study on heavy metal distribution along the lower stretch of Ganga (Hooghly) river, India. Department of Earth Sciences, Indian Institute of Science Education and ….
19. Mondal P, Nandan A, Siddiqui NA, Yadav BP (2014) Impact of soak pit on groundwater table. Environ Pollut Control J 18:12–17
20. Mukhopadhyay G, Rajeevan R, Archive B (2018) A review on microbial characteristic in ganges water, 9(3):10–20

21. Nandan A, Siddiqui N, Singh C, Aeri A, Gwenzi W, Ighalo JO, de Carvalho Nagliate P, Meili L, Singh P, Chaukura N (2021) COVID-19 pandemic in Uttarakhand, India: environmental recovery or degradation? 9(6):106595
22. Nandan A, Yadav BP, Baksi S, Bose D (2017) Assessment of water footprint in paper & pulp industry & its impact on sustainability. World Sci News 64:84–98
23. Ora A, Nandan A, Kumar A (2017) Hazard identification of chemical mixing plant through hazop study. Int J Advan Res Dev 2(3)
24. Padgelwar S, Nandan A, Mishra AK (2021) Plastic waste management and current scenario in India: a review. Int J Environ Anal Chem 101(13):1894–1906
25. Panwar S, Gaur D, Chakrapani GJ (2017) Total organic carbon transport by the Alaknanda River, Garhwal Himalayas, India. Arab J Geosci 10(9):207
26. Patil P, Sawant D, Deshmukh RJ IJOES (2012) Physico-chemical parameters for testing of water-a review, 3(3):1194
27. Rana D, Joshi N (2023) Assessment of human intervention on ecology of river—a case study. In: International conference on advances and innovations in recycling engineering. Springer, Singapore, pp 51–70
28. Repasch M, Scheingross JS, Hovius N, Lupker M, Wittmann H, Haghipour N, Gröcke DR, Orfeo O, Eglinton TI, Sachse D (2021) Fluvial organic carbon cycling regulated by sediment transit time and mineral protection, 14(11):842–848
29. Roy M, Shamim F, Pollution COI (2020) Research on the impact of industrial pollution on River Ganga: a review, 6(1):43–51
30. Santy S, Mujumdar P, Bala G (2020) Potential impacts of climate and land use change on the water quality of Ganga River around the industrialized Kanpur region, 10(1):1–13
31. Sati VP (2021) The environmental issues in Ganges Basin. Springer, The Ganges, pp 137–148
32. Sati VP (2021) Ganges. Springer
33. Saxena VJASRI., Impact Factor: 4.928 (2022) The study of physiochemical properties of Ganga's/tanneries water and their impact on crop productivity of plant, 8(1)
34. Sharma R, Kumar R, Sharma DK, Sarkar M, Mishra BK, Puri V, Priyadarshini I, Thong PH, Ngo PTT, Nhu V-H, Development and Sustainability (2021) Water pollution examination through quality analysis of different rivers: a case study in India, 1–22
35. Shukla AK, Ojha C, Shukla S, Garg R (2021) Water quality challenges in Ganga River Basin, India. In: The Ganga River Basin: a hydrometeorological approach. Springer, pp 1–19
36. Simon M, Joshi H (2022) Story of the Ganga River: Its pollution and rejuvenation. Riverine systems. Springer, 21–55
37. Singh P, Singh DS, Shukla UK (2018) Ganga: the Arterial River of India. The Indian Rivers. Springer, pp 75–92
38. Singh U, Singh S, Tiwari RK, RSJIAOAS, Pandey and Technology (2018) Water pollution due to discharge of industrial effluents with special reference to Uttar Pradesh, India—a review, 9(4):111–121
39. Subuddhi SP, Kansal A, Pandey P, Rana D, Ghoshal T. Study of impact of influx waste water generated at vehicle service centre on river water quality characteristics
40. Sun X, Fan D, Liu M, Liao H, Tian Y (2020) The fate of organic carbon burial in the river-dominated East China Sea: Evidence from sediment geochemical records of the last 70 years, 143:103999
41. Vijayaraman S, Mondal P, Nandan A, Siddiqui NA (2020) Presence of microplastic in water bodies and its impact on human health. In: Advances in air pollution profiling and control. Springer, Singapore, pp 57–65

Role of Earthworm (*Eisenia fetida*) in Bioconversion of Kitchen Waste in a World of Shifting Climatic Condition

Neha Tiwari, Damini Rana, and Namita Joshi

Abstract Urbanization, industrialization and agricultural activities accelerate the accumulation of solid waste around the globe, therefore becoming a major high tech and environmental problem. A sustainable approach to handle this waste is required. This paper highlights the recent trends of formulation of kitchen organic waste and its efficient management through vermicomposting. The vermicompost produced is of significant value along with excellent nutritional level, therefore regarded as a best method for waste disposal. Furthermore, it improves crop production and helps to attain long-term food security.

Keywords Earthworms · Vermicomposting · Kitchen organic waste · Waste disposal · Sustainable approach

1 Introduction

Earth's biological diversity altogether subsidize the universal productiveness of agricultural systems [10]. Soil flora and fauna form a major part of natural terrestrial ecosystems. Earthworms belonging to phylum Annelida account for a significant fraction of total biomass of invertebrates in soil and hence act as important decomposers [19]. Earthworms forms soil aggregates, thus greatly influencing soil structural characteristics [29].

In subtropical, tropical and temperate areas, earthworm represents nearly 80% of the total species diversity [39]. Earthworms are hermaphrodites and are considered as keystone species because they regulate soil fertility, water infiltration and soil

N. Tiwari (✉) · N. Joshi
Department of Environmental Sciences, Kanya Gurukula Campus, Gurukula Kangri (Deemed to Be University), Haridwar, Uttarakhand, India
e-mail: nemo.tiwari@gmail.com

D. Rana
Department of Environmental Sciences, Kanya Gurukula Campus, Gurukula Kangri (Deemed to be University), Haridwar & Central Laboratory, Uttarakhand Pollution Control Board, Dehradun, Uttarakhand, India

N. A. Siddiqui et al. (eds.), *Advances in Waste Management*, Lecture Notes in Civil Engineering 301, https://doi.org/10.1007/978-981-19-7506-6_9

detachability [30]. The contribution of the earthworms to the soil's quality has long been recognized and thus still regarded as important ecological regulators of the soil system [13].

The significant feature of earthworm lifecycle is their foraging behavior. Earthworm utilizes nourishment from the subsurface layer or plant debris, further grinding and expelling it, thus forming structures called cast [9]. Moreover, feeding and cast forming properties of the earthworm enhance the soil dynamics of terrestrial ecosystems and also manages the solid waste [14]. The earthworms are categorized into 3 different ecological category based on color, body morphology, burrowing ability as well as food preferences [32]. Earthworm acts as a natural bioreactor by transforming the biodegradable material into suitable bio fertilizers [38]. Nowadays, application of the earthworm for efficient modification of bio-waste towards high-quality compost is gaining popularity in many nations throughout the world, as well as in India [49].

Epigeic earthworm directly feeds on the substrate, thereafter strongly affecting the decomposition processes [33]. Various anthropogenic activities produce huge quantities of solid waste [42] all over the world, furthermore becoming a prominent environmental and technological problem [59]. The global scenario for solid waste expansion is deteriorating and is expected to rise to 3 billion tonnes by the year 2025 [8]. These wastes are extremely contagious because they include a variety of harmful microbes [36].

According to [50] the urban garbage alters the soil dynamics such as pH, electrical conductivity, water holding capacity, organic carbon content and bulk density. Hence, different approaches for proper solid waste management have been proposed and implemented all around the world [7]. Subsequently, vermicomposting is regarded as a viable technology for recycling the waste material into eco-friendly manure via economically and ecologically sound method. Vermicomposting is an aerobic, bio oxidative process that varies substantially from traditional methods as the organic material is metabolized by the gastrointestinal system of the earthworm [47].

Earthworms are the primary tools for bioconversion of solid organic waste and *Eisenia fetida* is considered as a most effective species for profitable use in the field of vermitechnology. It can consume around half of its body weight in a single day [28]. End-product of the vermicomposting is vermicompost, a nutritive 'organic fertilizer' rich in NPK and contains worm mucus that prevents the loss of nutrients [44]. The microbial decomposer community and earthworms exhibit a strong relationship in waste hierarchy, indeed earthworms support and stimulate both chemical as well as biological disintegration that further accelerate the activity of microbial communities [58].

Due to urbanization and industrialization, the situation has become alarming and two major problems have been raised *i.e.,* waste generation and global warming [41]. For the effective management of solid waste, the vermicomposting method has been used widely. Taking all this into account, this study was undertaken to check the efficiency of the earthworm (*Eisenia fetida*) in bioconversion of municipal solid waste (kitchen waste) into vermicompost. And also, the nutrient content of the produced vermicompost is determined for its effective use as an organic fertilizer to enrich the soil quality and fertility.

2 Materials and Method

2.1 Description of Study Area

For the present study, the substrate (kitchen waste) was collected from Jagjeetpur, Haridwar. The geo-coordinates of Haridwar is 29°94′ latitude and 78°16′ longitude with a subtropical climate (Fig. 1). The kitchen waste was mixed individually with cow dung and soil. Prior to this, the kitchen waste and cow dung used were dried before use.

2.2 Import of Earthworms

The vermiworms (*Eisenia fetida*) were initially procured from the garden area of Kanya Gurukul Campus, Haridwar (Fig. 4). Cemented pots of dimensions 39.5 cm

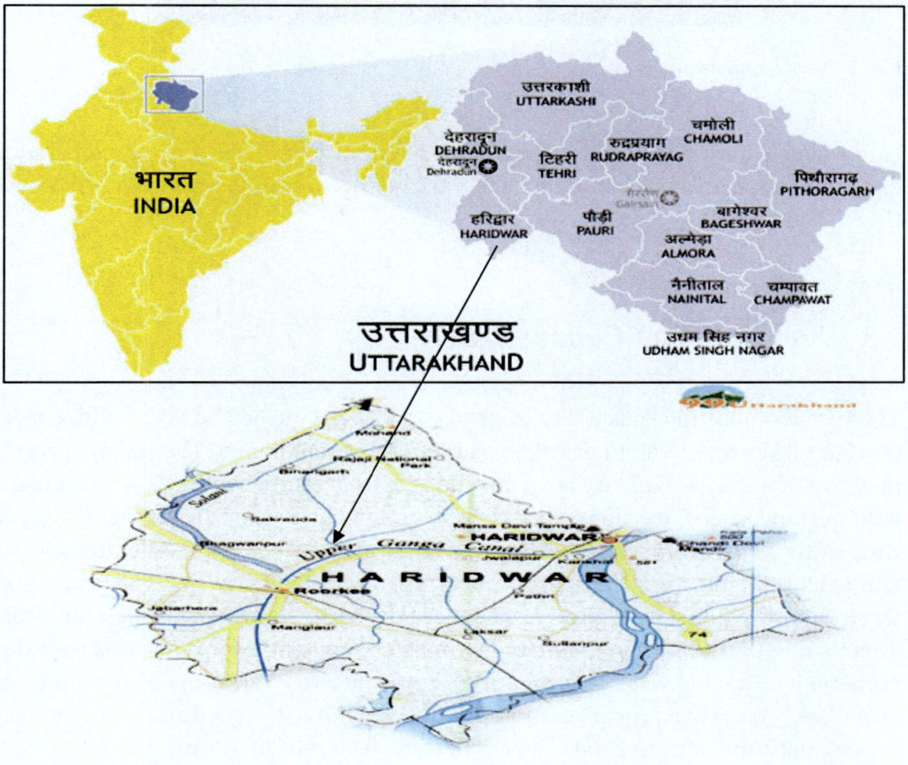

Fig. 1 Study area

Fig. 2 Vermicompost

height and 28 cm wide, with a hole at the bottom were used for the preparation of vermicompost. About 40 clitellated earthworms were inoculated into the beds (Figs. 2 and 3).

2.3 Preparation of Culture Bed

The vermicomposting beds were designed as per [24]. Crushed bricks and thick layer of sand makes up the first layer that ensures efficient drainage. The second layer is made up of loamy soil of height 15 cm and is appropriately hydrated and inoculated with vermiworms. Cow dung was strewn throughout the topsoil in the 3rd layer. Rice straw of 10 cm thickness was used to cover the fourth layer in order to maintain temperature of the vermicomposting bed. The entire unit was secured with mesh fabric and stored in a dark place to safeguard the earthworms against pest and from direct sunlight. In order to maintain the moisture content, every unit was hydrated properly by spraying water twice a week. Temperature as well as pH were monitored on a weekly basis and proper aeration was done by turning the 3rd and 4th layer once a week until the vermicompost was collected. Watering of vermicomposting beds was discontinued 3 days prior to harvesting.

Fig. 3 Cast over the vermicomposting bed

Fig. 4 Clitellated *Eisenia fetida*

2.4 Physico-Chemical Parameters

During the vermicomposting process temperature, pH, humidity were recorded on a weekly basis. A portable composting thermometer was used to determine the temperature. Humidity was recorded by the compost moisture meter and pH was measured by the soil pH pen. After 60 days, the vermicompost was harvested, sieved using mess and air dried (Fig. 2). After 2 days of air drying, the collected vermicompost was subjected to physical and chemical analysis for evaluation of nutrient quality of vermicompost in order to check its suitability for agricultural use. Subsequently, pH, electrical conductivity, moisture content, potassium, organic carbon, nitrogen and phosphorus were tested. Moisture content was determined by the gravimetric method and pH was calculated by using the pH meter with a suspension of 1:5. Rapid titration method of [57] was used for measuring organic carbon. Total nitrogen was quantified by micro-Kjeldahl digestion method and spectrophotometer was used to assess the available phosphorus. Conductivity was determined by using the conductivity meter and potassium and calcium by the flame photometer technique.

2.5 Statistical Analysis

The data were given in the form of mean $\pm$ S.E. Excel and SPSS program version 19 were used for the statistical analysis of the data.

3 Results and Discussion

3.1 Pre-composting Process

The pre-composting process involves pH and temperature variation in the composting mixture.

3.1.1 Temperature

Temperature pattern of the kitchen waste upto 6 weeks can be observed in Fig. 5. The initial temperature of the vermicomposting unit was recorded as 30 °C [16] reported the similar initial temperature pattern. Regular temperature fluctuations occurred throughout 6 weeks of the composting period. During the 1st and 2nd week the value increased to 31 and 32 °C due to the microbial activity that causes heat generation by the waste as described by [6, 60]. Later it decreased to 29 and 27 °C in the 3rd and 4th week. According to [1, 3, 25, 40] the mesophilic bacteria

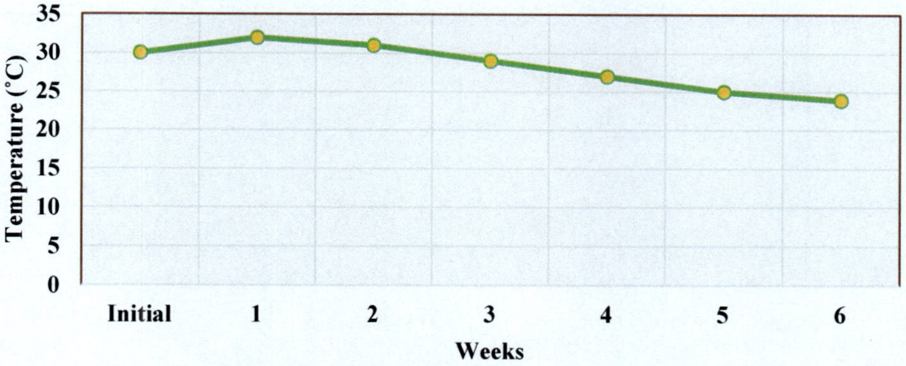

Fig. 5 Kitchen waste temperature behavior during the pre-composting process

become more active during the composting process, that becomes the key factor of temperature hike. During the last weeks the temperature of the waste started decreasing until its value approaches the ambient temperature clearly reflecting equilibrium, which itself, is consistent with the findings of [15, 35]. This stabilized temperature (25–23 °C) favors the growth, metabolic activity of earthworms and also favors the microbes associated with the earthworm.

3.1.2 pH

Figure 6 depicts the pH of kitchen waste during pre-composting, with the initial value of 8.3. In 1 stand 2nd week, the pH value shifted to 8.1 and 7.2 as a result of decomposition of organics and production of byproducts such as ammonium and humic acid as discussed by earlier studies of [2, 48]. Furthermore, the pH value becomes neutral (7) in the 3rd week, due to formation of CO_2 and the other organic acids. This was in line with the findings of [34, 37, 43, 45, 46, 52, 54]. Finally, in the last 3 weeks the pH value becomes acidic in nature, clearly showing the decreasing trend of pH with initially alkaline to finally acidic.

Solid waste management has become a serious environmental problem that the world is facing today [42, 53] and the household waste comprises 30–60% of biodegradable wastes. According to Food and Agriculture Organization of the United Nations (FAO), one-third including all foodstuff produced around the world is discarded [18]. Londhe and Bhosale considered waste as gold when it was transformed into vermicompost, thereafter concluded that solid waste was reduced by the earthworm. Vermicomposting is most efficient, viable and cost -effective method to decompose organic fractions of solid waste in an environment friendly way and is applicable in farming activities without causing any damage.

After 60 days of the decomposition process, the vermicompost had a blackish brown color, fairly similar to that as documented by [56]. The temperature was measured weekly and recorded as 27 °C (mean). Many studies suggested 20–35 °C

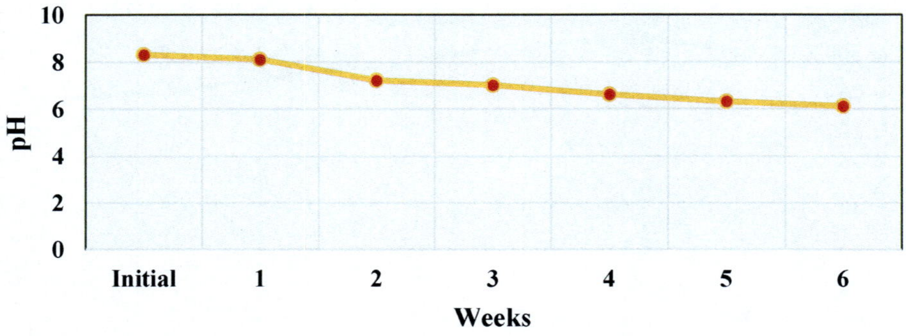

Fig. 6 Kitchen waste pH behavior during the pre-composting process

was the apt temperature to facilitate the production of vermicompost. The humidity of the vermicomposting bed was in the range of 65–75%.

The mean pH was 7.3 that falls under the normal range of 5–9 as per given by [12]. Also, according to [21], the vermicompost having pH level near neutral proves beneficial for the remediation of acidic soils. The results obtained are elucidated in Table 1 and the main elements namely nitrogen, phosphorus and potassium were 2.01%, 0.63% and 1.05% respectively. The nutrient status of the formed vermicompost is corroborated with the findings of [4, 5, 23]. According to [20, 26] microfloral population found in the gut of *Eisenia fetida* increased the K content by 10%. However, vermicompost contains phosphorus- solubilizing microbes that convert P into an easily available form [11, 31].

Moisture content of the vermicompost was found to be 68.3%. The study of [17] revealed that *Eisenia fetida* requires high moisture content for proper growth and development. Furthermore, [60] bring out that the proportion of water content can vary depending on the nature of substrate chosen for the decomposition process. The

Table 1 Physico-chemical characteristics of vermicompost prepared from the kitchen waste (All values are mean ± S.E)

S. No	Parameters (Unit)	Vermicompost
1	Moisture content (%)	68.3 ± 0.204
2	pH	7.39 ± 0.45
3	Conductivity (Scm^{-1})	1.93 ± 0.05
4	Phosphorus (%)	0.63 ± 0.606
5	Carbon content (%)	26.2 ± 0.430
6	Nitrogen (%)	2.01 ± 1.073
7	Potassium (%)	1.05 ± 0.501
8	Magnesium (%)	1.52 ± 0.19
9	Calcium (%)	1.24 ± 0.65
10	C/N ratio	13:1

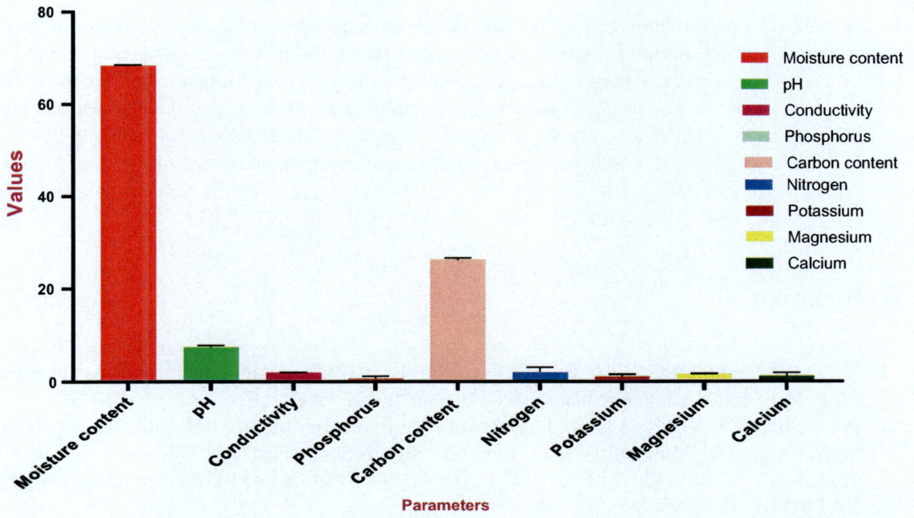

Fig. 7 Graphical representation of values of vermicompost

soluble salt concentration in the sample was 1.93 Scm^{-1}. The value of electrical conductivity (EC) was similar to the study of [55].

Organic carbon was reported to make up 26.2% of the total, which was significantly higher [51] illuminate that carbon is the building block of an organism and acts as the source of energy in the process of vermicomposting. Moreover, values of calcium and magnesium were 1.24 and 1.52, respectively, as seen in Fig. 7. The findings of [16] stated that C/N ratio of less than 20 is a prerequisite criterion for the compost maturity. During the study C/N ratio was recorded as 13.1 that was consistent with the findings of [22]. Reference [27] conclude that vermicomposting means converting waste into wealth and vermicompost helps to develop sustainable agriculture thus balancing the ecosystem.

4 Conclusion

The study was undertaken for evaluating the potential of *Eisenia fetida* in transforming organic waste into usable vermicompost by examination of the overall nutrient status of the finished product. The produced vermicompost had a dark color and all the essential nutrients. Therefore, the study recommends the utilization of earthworm species in the regulation of solid waste, as vermitechnology is a low-cost, environment friendly and sustainable technique for solid waste disposal. Furthermore, the chemical composition of the vermicompost is to accomplish the nutritional needs of plants, thus replacing the chemical fertilizer. It is also important to encourage more research on vermicompost methodology.

Disclaimer The presentation of material and details in maps used in this chapter does not imply the expression of any opinion whatsoever on the part of the Publisher or Author concerning the legal status of any country, area or territory or of its authorities, or concerning the delimitation of its borders. The depiction and use of boundaries, geographic names and related data shown on maps and included in lists, tables, documents, and databases in this chapter are not warranted to be error free nor do they necessarily imply official endorsement or acceptance by the Publisher or Author.

References

1. Acosta DCM, Solis PO, Villegas TOG, Cardoso VL (2013) Precomposteo de residuos organicos y su efecto enla dinamica poblacional de Eisenia foetida. Agron Costarric 37:127–139
2. Ali U, Sajid N, Khalid A, Riaz L, Rabbani MM, Syed JH, Malika RN (2015) A review on vermicomposting of organic wastes. Environ Progress Sustain Energy, 1–13
3. Ansari AA, Hanief A (2015) Microbial degradation of organic waste through vermicomposting. Int J Sust Agric R 2(2):45–54
4. Ansari A, Sukhraj K (2010) Effect of vermiwash and vermicompost on soil parameters and productivity of okra (*Abelmoschus esculentus*) in Guyana. Afr J Agric Res 5(14):1794–1798. https://doi.org/10.5897/ajar09.107
5. Ansari AA, Jaikishun S, Islam SK, Kuri KF, Nandwani D (2016) Principles of vermitechnology in sustainable organic farming with special reference to Bangladesh. In: Nandwani D (eds) Organic farming for sustainable agriculture. Sustainable development and biodiversity. 9. Springer International Publishing, Switzerland, pp 213–229. https://doi.org/10.1007/978-3-319-26803-3_10
6. Bueno MP, Diaz BMJ, Cabrera CF (2008) Capitulo 4. Factores que Afectan al Proceso de Compostaje
7. Chang NB, Davila E (2008) Municipal solid waste characterizations and management strategies for the Lower Rio Grande Valley. Texas. Waste Manag 28(5):776–794
8. Charles W, Walker L, Cord-Ruwisch R (2009) Effect of pre-aeration and inoculum on the start-up of batch thermophilic anaerobic digestion of municipal solid waste. Bioresoure Technol 100:2329–2335
9. Curry J (1998) Factors affecting the abundance of earthworms in soils. In: Earthworm Ecology. https://doi.org/10.1201/9781420039719.pt3
10. Decaens T (2010) Macroecological patterns in soil communities. Glob Ecol Biogeogr 19:287–302
11. Dohaish JAB (2020) Vermicomposting of organic waste with *Eisenia fetida* increases the content of exchangeable nutrients in soil. Pak J Biol Sci 23(4):501–509. https://doi.org/10.3923/pjbs.2020.501.509
12. Dominguez J, Edwards CA (2011) Biology and ecology of earthworm species used for vermicomposting. In: Edwards CA, Arancon NQ, Sherman (eds) Vermiculture technology: earthworms, organic wastes, and environmental management. CRC Press, Taylor & Francis Group, Boca Raton, USA, pp 27–40
13. Edwards CA, Bohlen PJ (1996) Earthworm ecology and biology, 3rd edn. Chapman & Hall, London, pp 196–212
14. Edwards WM, Shipitalo MJ (1998) Consequences of earthworms in agricultural soils: aggregation and porosity. In: Earthworm Ecology. St. Lucie Press, Boca Raton, Florida
15. FAO (2013) Manual de Compostaje del Agricultor: Experiencias en Am*er*ica Latina; Organization de las Naciones Unidas para la Alimentacion y la Agricultura: Santiago, Chile. ISBN 978-92-5-307844-8
16. Gonzalez EC, Diaz MRG, Salas LDM, Castillo MPS (2019) Pre-Composting and vermicomposting of pineapple (*Ananas Comosus*) and vegetable waste. Appl Sci 9:3564. https://doi.org/10.3390/app9173564

17. Gunadi B, Edwards CA (2003) The effects of multiple applications of different organic wastes on the growth, fecundity and survival of *Eisenia fetida* (Savigny) (Lumbricidae). Pedobiologia 47:321–329
18. Gustavsson J, Cederberg C, Sonesson U (2011) Global food losses and food waste: extent, causes and prevention. Food and Agriculture Organization of the United Nations (FAO), Rome, pp 1–29
19. Haokip SL, Singh TB (2012) Diversity and distribution of earthworms in a natural reserved and disturbed sub-tropical forest ecosystem of Imphal-West, Manipur. India. Int Multidisciplinary Res J 2(2):28–34
20. Hashimoto S, Furuya M, You X, Wanibuchi G, Tokumoto H, Tojo M, Shiragaki K (2021) Chemical and microbiological evaluation of Vermicompost made from school food waste in Japan. JARQ. 55(3):225–232
21. Huang K, Li F, Wei Y, Fu X, Chen X (2014) Effects of earthworms on physicochemical properties and microbial profiles during vermicomposting of fresh fruit and vegetable wastes. Bioresoure Technol 170:45–52. https://doi.org/10.1016/j.biortech.2014.07.058
22. Huntley S, Ansari AA (2021) Vermicomposting evaluation of different combinations of organic waste using *Perionyx excavates*. Int J Recycling Organic Waste Agric 10:287–295
23. Ismail SA (1997) Vermicology: the biology of earthworms. Orient Longman Press, Hyderabad, p 92
24. Ismail SA (2005) In: The earthworm book. Other India Press, Mapusa, p 101
25. Jara-Samaniego J, Perez-Murcia MD, Bustamante MA, Paredes C, Perez-Espinosa A, Gavilanes-Teran I, Lopez M, Marhuenda-Egea FC, Brito H, Moral R (2017) Development of organic fertilizers from food market waste and urban gardening by composting in Ecuador. PLoS ONE 12(7):e0181621
26. Kaviraj, Sharma S (2003) Municipal solid waste management through vermicomposting employing exotic and local species of earthworms. Bioresoure Technol 90:169–173
27. Korav S, Malannavar AB, Sharma L (2021) A review-vermicomposting: an effective option for agriculture waste management. Biolog Forum Int J 13(2):211–219
28. Kumar A, Prakash CHB, Brar NS, Kumar B (2018) Potential of vermicompost for sustainable crop production and soil health improvement in different cropping systems. Int J Curr Microbiol Appl Sci 7:1042–1055
29. Lavelle P (1997) Faunal activities and soil processes: adaptive strategies that determine ecosystem function. Adv Ecol Res 27:93–132
30. Lavelle P, Spain AV (2001) Soil ecology. Kluwer Academic Publishers, New York. https://doi.org/10.1007/978-94-017-5279-4
31. Lee KE (1985) Earthworms: their ecology and relationships with soils and land use. Academic Press, Sydney, p 411
32. Lemtiri A, Colinet G, Alabi T, Cluzeau D, Zirbes L, Haubruge E, Francis F (2014) Impacts of earthworms on soil components and dynamics. a review. Biotechnol. Agron Soc Environ 18(1):121–133
33. Londhe PB, Bhosale SM (2015) Recycling of solid wastes into organic fertilizers using low cost treatment: vermicomposting. Int J Innov Eng Res Technol 2:1–11. https://goo.gl/0FouGw
34. Lopes FL, Lopes SWT, Milori DMBP, Luiz SM, Martin NL (2010) Characterization of organic matter from composting of different residues by physicochemical and spectroscopic methods. Bioresoure Technol 101:1927–1936
35. Mondal P, Nandan A, Siddiqui NA, Yadav BP (2014a) Impact of soak pit on groundwater table. Environ Pollut Control J 18(1):12–17
36. Mondal P, Nandan A, Siddiqui DN, Yadav BP (2014b) Removal of fluoride from water by suitable low cost environmental friendly methods. Environ Pollut Control J 17(6)
37. Nagavallemma KP, Wani SP, Lacroix S, Padmaja VV, Vineela C, Babu RM, Sahrawat KL (2006) Vermicomposting: recycling wastes into valuable organic fertilizer. Global theme on agroecosystems, Report No. 8. Int Crops Res Inst Semi-Arid Tropics 2(1):1–20
38. Nainawat R, Bhardwaj N (2001) Density and distribution of earthworms in different localities of Jaipur. J Eco Physiol 4(1/2):9–13

39. Nair J, Sekiozoic V, Anda M (2005) Effect of pre-composting on vermicomposting of kitchen waste. Bioresoure Technol 97:2091–2095
40. Nandan A, Siddiqui NA, Singh C, Aeri A (2021) Occupational and environmental impacts of indoor air pollutant for different occupancy: a review. Toxicol Environ Heal Sci 13(4):303–322
41. Nandan A, Yadav BP, Baksi S, Bose D (2017) Recent scenario of solid waste management in India. World Scientific News 66:56–74
42. Narayan J (2000)Vermicomposting of biodegradable wastes collected from Kuvempu University campus using local and exotic species of earthworm. In: Proceedings of a national conference on industry and environment, Karad, India, pp 417–419
43. Oluseyi EE, Ewemoje TA, Adedeji AA (2016) Comparative analysis of pit composting and vermicomposting in a tropical environment. Int J Biolog Biomolec Agric Food Biotechnol Eng 10(3):174–177
44. Pagaria P, Totwat KL (2007) Effects of press mud and spent wash in integration with s with phosphogypsum on metallic cation buildup in the calcareous sodic soils. J Indian Soc Soil Sci 55(1):52–57
45. Panday SN, Yadav A (2009) Effect of vermicompost amended alluvial soil on growth and metabolic responses of rice (*Oryza sativa* L.) plants. J Eco-friendly Agric 4(1):35–37
46. Pathma J, Sakthivel N (2012). Microbial diversity of vermicompost bacteria that exhibit useful agricultural traits and waste management potential. Springer Plus 1:26. https://goo.gl/pv6t39
47. Pramanik P, Ghosh G, Ghosal P, Banik P (2007) Changes in organic–C, N, P and K and enzyme activities in vermicompost of biodegradable organic wastes underliming and microbial inoculants. Bioresoure Technol 98:2485–2494
48. Rajkhowa DJ, Bhattacharyya N, Sarma VK, Mahanta K (2014) Diversity and distribution of earthworms in different soil habitats of Assam, North-East India, an Indo-Burma Biodiversity Hotspot. Proc Nat Acad Sci India Sect B Biol Sci 85(2):1–10. https://doi.org/10.1007/s40011-014-0380-1
49. Rakkini VM, Vincent S, Kumar AS, Baskar K (2017) An overview: organic waste management by earthworm. J Civil Eng Environ Sci 3(1):013–017
50. Ramnarain YI, Ansari AA, Ori L (2019) Vermicomposting of different organic materials using the epigeic earthworm *Eisenia fetida*. Int J Recycling Organic Waste Agric 8:23–26
51. Singh J (1997) Habitat preferences of selected Indian earthworm species and their efficiency in reduction of organic material. Soil Biol Biochem 29:585–588
52. Sumi MG, Vani M, Idicula DV, Mini KD (2014) Solid waste management using vermicomposting and kodmic? Bio-pedestal column and its utility as organic manure. Asian J Microbiol Biotechnol Environ Sci 16:333–338. https://goo.gl/A0NulQ
53. Suthar S (2008) Development of a novel epigeic-anecic-based polyculture vermireactor for efficient treatment of municipal sewage water sludge. Int J Environ Waste Manag 2 (½): 84–101
54. Tiwari N, Negi N, Raj R, Joshi N (2017) A study on physico—chemical and myco—floral proportions of compost and vermicompost. J Env Bio-Sci 31(2):375–377
55. Vasanthi P, Kamala S, Balamurugan K, Udayakumar S (2018) Effect of Vermicomposting of Kitchen Waste Using Lampito mauritii. International J Res Analyt Rev (IJRAR) 5(4):515–519
56. Walkley A, Black CA (1934) An experiment of Degtjareff methods for determining soil organic matter and a proposed modification of the chronic acid titration methods. Soil Sci 37(1):29–38
57. Yadav A, Garg VK (2009) Feasibility of nutrient recovery from industrial sludge by vermicomposting technology. J Hazard Mater 168:262–268
58. Yadav A, Garg VK (2011) Recycling of organic wastes by employing *Eisenia fetida*. Bioresoure Technol 102:2874–2880
59. Yadav A, Garg VK (2019) Biotransformation of bakery industry sludge into valuable product using vermicomposting. Bioresoure Technol 274:512–517
60. Yadav J, Gupta KR (2017) Dynamics of nutrient profile during vermicomposting. Ecol Environ Conservat 23(1):516–521. http://www.envirobiotechjournals.com

Recent Trends of Temporal and Spatial Variation in Waste Generation and Its Impacts on River Water Quality: Special Emphasis on Suswa and Song Rivers of Uttarakhand

S. P. Subudhi, Ankur Kansal, Tarumay Ghoshal, Naveen Singhal, and Damini Rana

Abstract Survival and procreation of life on the planet is dependent on the water resources. Water being an unreplaceable vital resource sustains life of the species, ecological processes, agricultural and other developmental activities. A watershed is a natural structure for developing efficient utilization of precipitation that fluctuates spatially and temporally. Around 80% of untreated waste in the world is disposed from domestic and industrial sources which is creating a threat to living organisms. Different sampling sites were studied in Doon valley for a period of one year. Parameters chosen for the study are: Physicochemical parameters—pH, electrical conductivity, total hardness; Biological parameters—dissolved oxygen, biological oxygen demand; Bacterial pollution indicator parameters—total coliforms; Pesticides—Benzene hexachloride and endosulfan. According to the findings, there are significant deviations between different sites and parameters chosen. Water quality has been found to be unfit for use and extensively contaminated with various organic and inorganic pollutants at several locations. Effective management as well as awareness campaigns for water conservation are recommended.

Keywords Dehradun · Anthropogenic activities · pH · EC · Total hardness · DO · BOD · Total coliform · Pesticides · Suswa · Song · River pollution

---

S. P. Subudhi
DIT University, Dehradun and Uttarakhand Pollution Control Board, Dehradun, India

A. Kansal
Uttarakhand Pollution Control Board, Dehradun, Uttarakhand, India

T. Ghoshal · N. Singhal
DIT University, Dehradun, Uttarakhand, India

D. Rana (✉)
Department of Environmental Sciences, Gurukula Kangri (Deemed to be University), Haridwar, Uttarakhand, India
e-mail: daminiranaofficial@gmail.com

Central Laboratory, Uttarakhand Pollution Control Board, Dehradun, India

N. A. Siddiqui et al. (eds.), *Advances in Waste Management*, Lecture Notes in Civil Engineering 301, https://doi.org/10.1007/978-981-19-7506-6_10

1 Introduction

Water is essential for all living beings including human society underpinning all aspects from ecology, agriculture to industry, and it cannot be replaced by any alternative resource. A river basin is a natural entity for planning productive uses of available rainwater, which varies in a spatial and temporal manner (Srikantaswamy). It has been estimated that around 80% of all the industrial and municipal wastewater is released into the environment, without being treated and posing threat on human health and ecosystems around the world. Disposal of the waste including hazardous waste also needs to be focused on with the better strategies, models and techniques [24, 28]. In least developed countries, where sanitation and wastewater treatment facilities are severely insufficient, this ratio is substantially greater [9].

River water quality is easily deteriorated and over exploited being exposed to interactions with anthropogenic & environmental variables unlike ground water [6]. Clean and safe natural sources of water are short-lived in an ecosystem due to various environmental and anthropogenic factors influencing its physico-chemical attributes in space and time [31]. With rapid increase in urbanization, industrialization and other infrastructural developments, domestic waste along with industrial effluent is being disposed off from various points and the non-point sources into the river [17]. Dumpsites grow in number and size over time due to population turnover [23].

Excess nutrients in agricultural runoff are also one of the most common water quality-related concerns around the world [22]. The pollutants from these sources increase the pollution load in the river thus affecting the physico-chemical and biological characteristics [27]. Many sacred rivers and springs originate in Uttarakhand, India, and they play vital roles in Indian culture and history. However, residential garbage, rock weathering, anthropogenic activities, and sewage effluents, pollute the rivers of Uttarakhand, affecting their physico-chemical and biological properties [28, 39], which directly relates the river water quality [30].

According to a report submitted to the Central Pollution Control Board (CPCB) of India by the Uttarakhand Environment Protection and Pollution Control Board (UEPPCB 2007), effluents from paper industries discharge nearly 141,620 Kilo Litres per Day (KLD) effluent, discharge from sugar industries nearly 24,137 KLD, and discharge from distilleries nearly 6,000 KLD, into Uttarakhand river bodies [11]. Aside from that, Uttarakhand Pey Jal Nigam (UPJN), 2009, mentioned that 97.37 million of Litres per Day (MLD) sewage, whether treated or untreated, finds its way into Uttarakhand's water bodies. Recent data collected from Uttarakhand Pollution Control Board (UKPCB), 2021 gives the overall status of Uttarakhand with an estimated population of 1.14 Crore in 2021 having Urban Sewage generation of about 329.32 MLD and Rural Sewage generation of about 225.18 MLD.

Water quality and quantity deteriorate as a result of water consumption for household usage, agricultural output, industrial production, mining, power generation, forestry operations, etc. [15] which can be supported by the study of [19] showing rivers during lockdown period were also found to have better water quality status indicating less anthropogenic activities and thus lower quantity of impurities in the water [21]. According to Census of India [38] Population projection of Uttarakhand (2019) depicts the increasing trend of population from year 2011–2021 which may lead to various socio-economic problems in the state. The increase in unplanned settlement either directly or indirectly are impacting the hilly regions having vast ecological values [8]; hence, it has becomes a major concern for a city like Dehradun in Uttarakhand in India [10]. As this region comes under hilly terrain, thus natural slope and its stability factor become its essential aspects of the perception of the land use. The variations of the land use-land cover from 1972 to 2016 in Dehradun is presented by [7] showing that the built-up area has increased continuously with decrease in agricultural land since year 1972 to 2016. Slight decrease in forest land, vegetation and water bodies has also been observed from year 1972–2016 while almost no change has been observed in plantation and scrub forest in these years.

Dehradun is capital city of the Himalayan state, Uttarakhand and is known for its scenic beauty and also having an important place in Hindu mythology. It is one among the oldest cities in our country India. It's an attractive place with quaint landscape (Yadav, Nandan et al.), delightful climate and receives an average annual rainfall of 2073.3 mm [34].

Dehradun and Doiwala are two important urban settlements in the Suswa and Song river's watershed. Dehradun has a population of 578,420 people (2011 Census), whereas Doiwala has a population of 870. The rivers Suswa and Song are two of the Ganga's principal tributaries. Along the Suswa River, there are 11 settlements in addition to the urban area. The Rispana and Bindal rivers are two main drainage networks that convey Dehradun's urban drainage and join the Suswa River at Mothrawala(Garg). The Suswa River drains the eastern half of Dehradun and merges with the Song River before flowing into the Ganga. River Suswa joins various drainage, including River Song, after confluence with Rivers Rispana and Bindal near Mothrawala [4]. This catchment receives water from 51 drains, including the river Song.

CPCB has designated the region of Mothrawala to Raiwala as a contaminated river stretch based on water quality data from 2016 and 2017. River Song is known to be spring fed river that is originating from different small rivulets of the mountainous range of Dhanolti forest and crossing with the Sahastradhara streams it is flowing downward towards the basins of Doon valley and then finally it assimilates into river Ganga at Raiwala. Song river flows along Raiwala, Doiwala, Chiddarwala, and Lacchiwala and is the ultimate source of water in these regions [14]. Song basin carries effluent from the distilleries and sugar mills along its course. This work aims to find out the possible factors responsible for degradation of rivers, which are the only source of drainage in Dehradun, Uttarakhand.

2 Materials and Methods

The work was carried out for duration of one year from January 2021-December 2021 on monthly basis. Water quality was assessed for four seasons Winter (January and February), Summer (March, April and May), Monsoon (June, July, August and September) and Post-Monsoon (October, November and December). Seasonal relationship was also assessed to know effects of different environmental conditions on the river water. Collection of the water samples was done in sterilized sampling bottles. Five major sites along the rivers Suswa and Song (S1, S2, SN1, SN2 and G1) and eight important parameters including physico-chemical, biological and pesticides were chosen for the study. Figure 1 describes the Sampling Location Map of the study area.

Physico-chemical parameters like Dissolved oxygen (DO) in mg/l was performed on the site itself and other remaining parameters like Hydrogen potential (pH), Electrical conductivity-(EC) in μmho/cm, Total hardness-(TH) signifying Calcium and Magnesium hardness in mg/l, Biological oxygen demand-(BOD) in mg/l and bacteriological parameter- Total coliform-(TC) were analyzed in laboratory by Most Probable Number (MPN). pH and EC were measured by using digital meter named pH meter and Conductivity meter, respectively. TH, DO and BOD were analyzsed by titration method. Pesticides like Benzene hexachloride (BHC) and Endosulfan (ES) were analyzed by using Gas chromatography. Standard Methods [1] were followed for analysis of water samples. The Table 1 gives the information regarding codes, coordinates and locations of sampling sites.

3 Results and Discussion

pH: pH is an important factor to test the chemistry of water as the biological makeup and state of various aquatic ecosystems are influenced by their pH levels [33]. Natural water has a pH range of 4 to 9, and most of it is slightly basic due to the presence of alkali and alkaline earth metal like carbonates and bicarbonates [2]. G1 with maximum fluctuation in values with a standard deviation of 0.85 is observed to have maximum value in summers while minimum in all the seasons. As shown in Fig. 2, site S1 has minimum fluctuation in pH values with standard deviation of 0.4 and highest value of pH among all other stations. Site S1 consists of river water mixed with sewage coming from the urban slums and site G1 is the downstream of Song which covers a long distance and due to dilution factor possess an average water quality.

EC: Electrical conductivity (EC) measures potential of a liquid to carry an electric current depending upon its ionic content. Water that is chemically pure doesn't conduct electricity. A larger concentration of dissolved inorganic materials in ionized state is indicated by high EC values. Pollution is indicated by any increase in the EC of water. Site S2 in summers, monsoon and post-monsoon has maximum EC

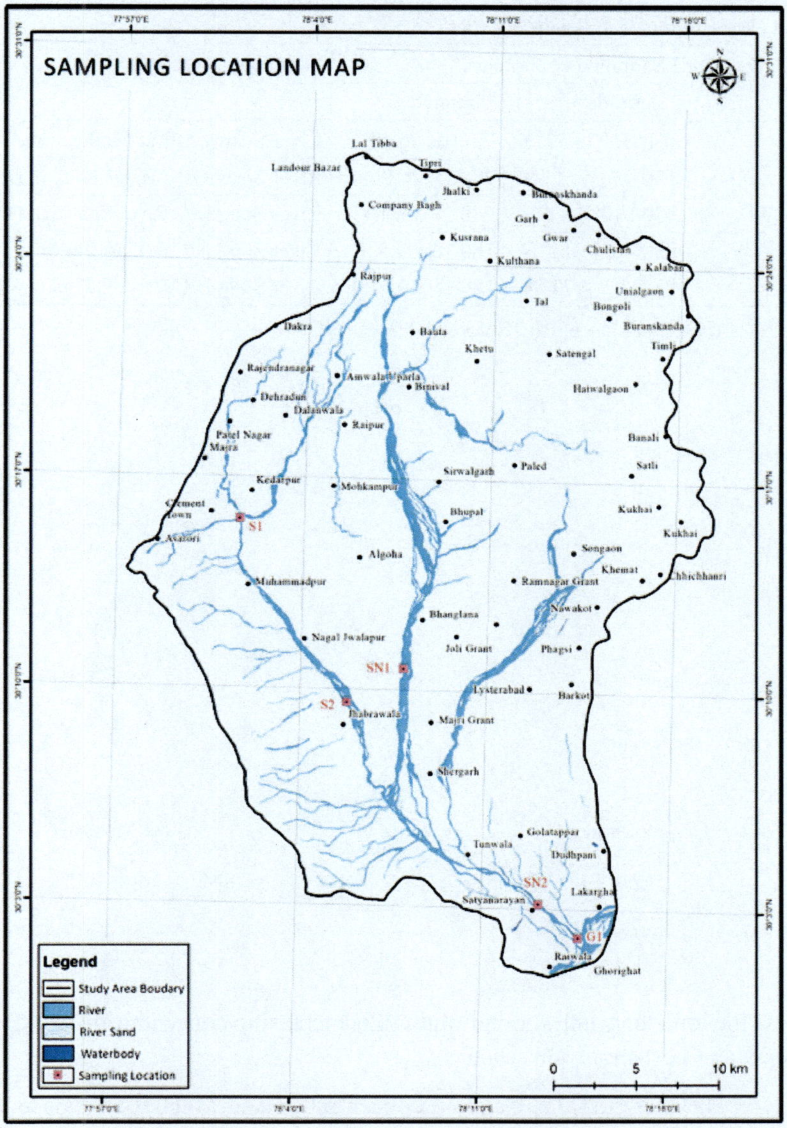

Fig. 1 Sampling location map

(Fig. 3). Site G1 in winters, summers, monsoon and post-monsoon has minimum EC. Maximum standard deviation of 211.3 is observed in site SN1 and minimum standard deviation of 17.5 is observed in site S1. The seasonal variation in surface water shows higher ionic concentration in the monsoon season and the post-monsoon season in comparison to winter and summer seasons. The mixing of water with various pollutants during monsoon season [35] and enhanced rock water interaction [5] probably

Table 1 Description of the sampling sites

Sampling codes	Sampling coordinates		Sampling sites
	Latitude	Longitude	
S1	30°15′ 33.52″N	78° 01′ 36.48″E	River Suswa A/C at Mothrowala
S2	30° 9′ 35.50″N	78° 05′ 46.96″E	River Suswa B/C River Song at Doiwala
SN1	300 10′ 42.2″N	780 07′ 53.1″E	River Song B/C River Suswa at Doiwala
SN2	300 03′ 10.2″N	780 13′ 08.5″E	River Song A/C at Doiwala
G1	30° 02′ 5.77″N	78° 14′ 39.44″E	River Song B/C River Ganga at Raiwala

* *Note* A/C-After Confluence; B/C-Before Confluence

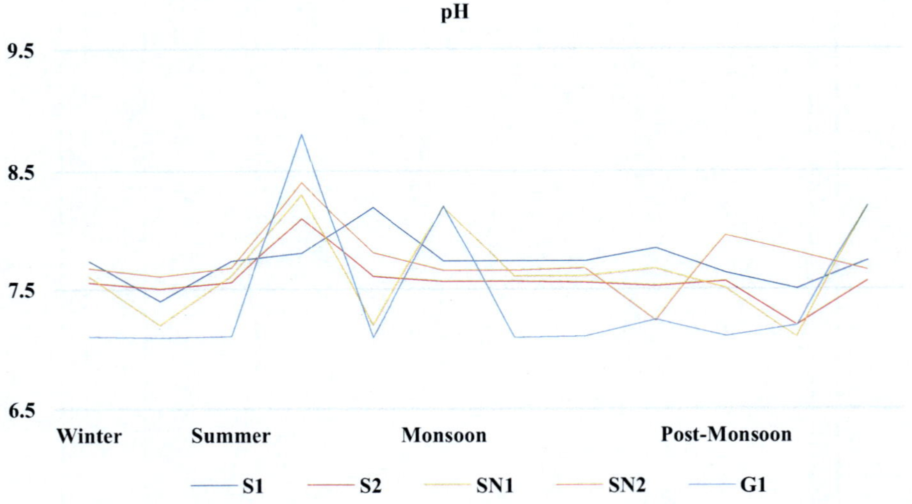

Fig. 2 pH

increase the ionic concentration in water thus increasing concentration of EC during monsoon and post-monsoon seasons.

TH: Total Hardness (TH) is Calcium concentration and Magnesium concentration expressed in the form of calcium carbonate in mg/l depending upon the subject of source. The divalent amount of calcium and magnesium, ferrous, manganese as well as strontium ions is referred to as water hardness. It primarily comes from rock and soil erosion. Site S2 in summers and monsoon has maximum hardness which may be due to increase in the evaporation rate resulting in high salt concentrations [5]. Site S2 also carries urban sewage waste (including slums) making it a polluted river stretch, resulting in increased hardness by depositing Ca^{2+} and Mg^{2+} salt concentrations [13]. As seen in Fig. 4, site G1 in winters has minimum hardness which can be due to dilution factor and higher flow rate as compared to other sites. Site SN1 has maximum standard deviation of 20.87 while site SN2 has minimum standard deviation of 2.31. Presence of carbonate rocks in Dehradun valley (Doon valley),

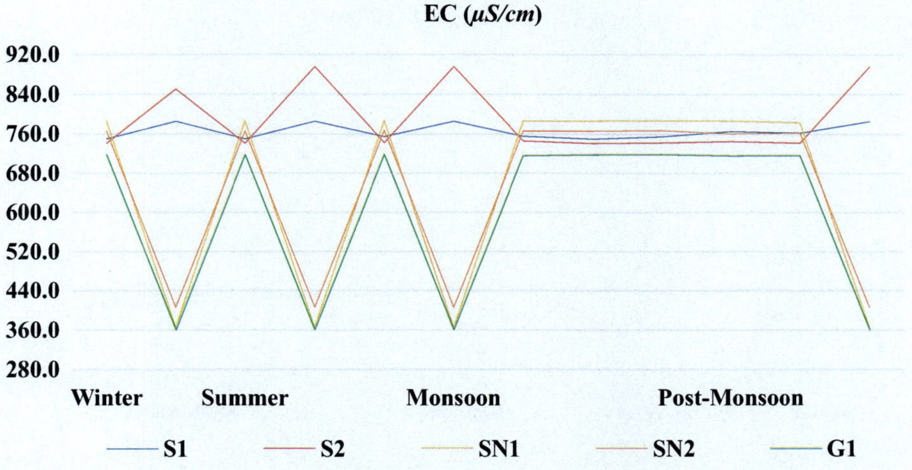

Fig. 3 Electrical conductivity

North region as the Krol formation (Limestone and Dolomite) may the reason behind higher concentration of TH in the river in all the sampling sites.

DO & BOD: Dissolved oxygen (DO) is considered as vital parameter which reflects physical as well as biological occurrences in by indicating the degree of pollution in water in mg/L [16]. The amount of dissolved oxygen in water decreases as oxygen-demanding substances, both organic or inorganic, are added to it. If the level of oxygen drops below the critical threshold, fish and other higher forms of aquatic life are at risk. Site G1 in winters, summers, monsoon and post-monsoon as observed in Fig. 5, has maximum DO which may be due to dilution factor and sufficient flow

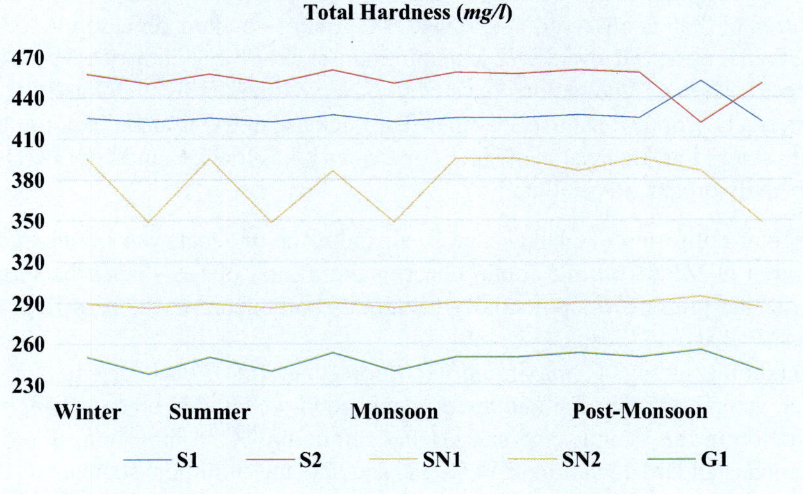

Fig. 4 Total haedness

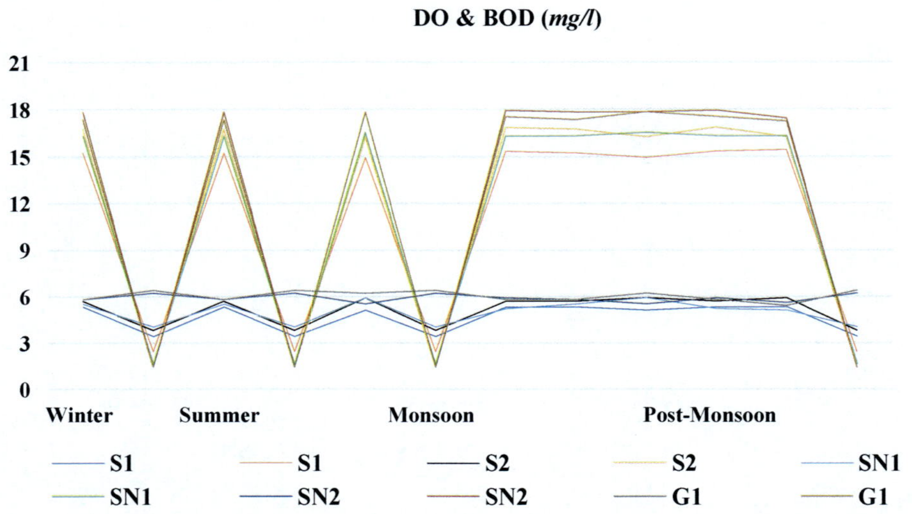

Fig. 5 Dissolved oxygen and Biological oxygen demand

rate of the river while site S1 in winters, summers, monsoon and post-monsoon has minimum DO which is found to be severely polluted stretch among all the locations.

Biological oxygen demand (BOD) is helpful in determining oxygen required in effluents as well as polluted waters in mg/L. BOD is the quantity of oxygen that living things (microbes) need to use or stabilize organic material over the course of five days. BOD readings are used to gauge the level of waste and water contamination since they roughly represent the quantity of oxidizable organic matter [33]. Site SN2 in monsoon and post-monsoon has maximum BOD and minimum BOD was observed in sites S2, SN2, G1 in winters, summers, monsoon and post-monsoon. Maximum standard deviation for DO of value 1.05 is observed in site S2 with minimum standard deviation of 0.35 is observed in site SN2. Maximum standard deviation for BOD of value 8.25 is observed in site SN2 with minimum standard deviation of 6.5 is observed in site S1. Highest fluctuations have been observed in site SN2. DO and the BOD are inversely proportional to each other. The decrease in DO is an indicator of higher BOD value. Thus, the results of site S1 indicates it has low DO and high BOD value and heavily organically polluted.

TC: Total coliforms (TC) presence is an indicator of fecal water contamination measured in MPN/100 ml. Total coliforms are a kind of rod-shaped bacteria that indicate the presence of potentially hazardous pathogenic bacteria in river water bodies (Pathak and [25].

According to Fig. 6, maximum TC (more than 1600) was seen in site S1 in winter, summers, monsoon and the post-monsoon which may be due to high slum population in the vicinity and site G1 has minimum TC in monsoon. Site S1 has maximum standard deviation of 14.72 and site SN2 has minimum standard deviation of value 2. Seasonal variations reveal the concentration of TC is highest in summers

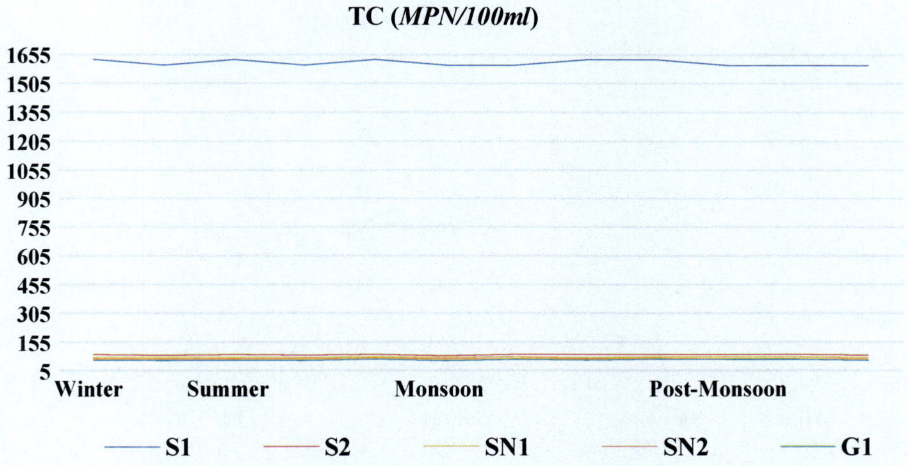

Fig. 6 Total coliform

which may be due to less river water and wide dry area near the river which in turn is exposed for various anthropogenic activities harmful for aquatic life [12]. High concentration of TC after summers is also visible during monsoon season which may be the post-effect of summer activities. In all the seasons Total Coliform is also found positively correlated with the Electrical conductivity and found negatively correlated with the Dissolved oxygen.

Pesticides: Pesticides pose substantial health risks to biological systems due to their quick response for being fat-soluble and bioaccumulate in creatures other than their intended targets. Pesticides can have a number of negative consequences, even at low concentrations. These effects can be observed at the biochemical, molecular, or behavioral levels. Drainage, rainfall, microbiological activity, soil temperature and application rate, the mobility, solubility as well as half-life of pesticides are elements that might cause pesticides and their residues to pollute water [3].

Both the pesticides tested Benzene hexachloride (BHC) and Endosulfan (ES) are in the range of 0–0.09 μg/L. Pesticides positively correlated with each other in all seasons except in Winters. According to Fig. 7, maximum BHC was observed in site SN1 in summers and site SN2 in winter, summers, monsoon and post-monsoon; BHC was not detected in February (winters) and August (monsoon). ES was observed in summers, monsoon and post-monsoon in all points, where site SN2 in the post-monsoon has maximum ES and not detected in winters and in the month of August (monsoon). No deviation was observed in most of the points for both the pesticides except the maximum value of 0.00176 which was observed at site SN2 for ES. Presence of low content of these pesticides may be due to alternative pesticides used for the agricultural fields in the nearby regions also supported by [37].

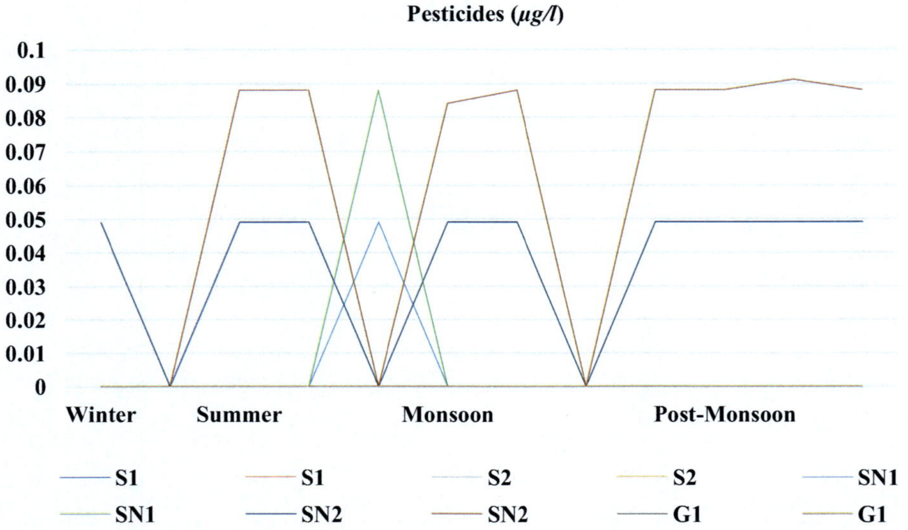

Fig. 7 Pesticides

4 Conclusion

Study conducted on Suswa and Song rivers revealed that there are various deviations in different parameters with respect to sites. Temporal and seasonal variations can be clearly observed in the study locations. Quality of water at various locations is found unfit for use and heavily contaminated with various organic as well as inorganic pollutants which may be due to urbanization, improper waste management, sewage discharge [20], and fecal discharge in and nearby riparian zones.

It is concluded that the rivers of Doon valley are contaminated with municipal waste, fecal waste, plastics waste [36] and wastewater effluent [32]. Difficulties are faced in quantifying household waste discharge and designing specific ways of promoting compost and recycling, especially in developing nations with limited resources [26]. Understanding and quantifying sources of pollution in rivers, as well as spatial and temporal variability, are required to implement mitigation solutions. Strategies like pollutant source reduction, mitigation, management of the river ecology and environmental education are recommended for effective regional litter management. Only cleaning up the waste generated does not prevent the opportunistic dumping thus, in order to stop these practices different stakeholders and the community must work together.

References

1. American Public Health Association, A. W. W. A. a. W. E. F (2017) Standard methods for the examination of water and wastewater. American Public Health Association, Washington, DC 23
2. Amor HB, Elaoud A, Hozayn M (2018) Does magnetic field change water pH? 8(1):1–7
3. Anju A, Ravi S P, Bechan S (2010) Water pollution with special reference to pesticide contamination in India
4. Bahukandi KD, Ahuja NJ (2013) Building block methodology assisted knowledge-based system for environmental-flow assessment of Suswa River of Dehradun Dist., India: a reminiscent framework2(12):74–80
5. Bartarya SK, Bahukhandi DK (2012) Impact assessment of urbanization and industrialization on surface and groundwater quality1(1):11–22
6. Bhutiani R, Khanna DR, Kulkarni DB, Ruhela M (2016) Assessment of Ganga river ecosystem at Haridwar, Uttarakhand, India with reference to water quality indices. Appl Water Sci 6(2):107–113
7. Dey J, Sakhre S, Gupta V, Vijay R, Pathak S, Biniwale R, Kumar R (2018) Geospatial assessment of tourism impact on land environment of Dehradun, Uttarakhand, India. Environ Monit Assess 190(4):181
8. Estoque RC, Murayama YJL, u. planning (2013) Landscape pattern and ecosystem service value changes: Implications for environmental sustainability planning for the rapidly urbanizing summer capital of the Philippines 116:60–72
9. Fayomi G, Mini S, Fayomi O, Owodolu T, Ayoola A, Wusu O (2019) A mini review on the impact of sewage disposal on environment and ecosystem. IOP conference series: earth and environmental science, IOP Publishing
10. Kishore MV, Garg A (2011) Issues and solutions of drainage system for Dehradun City 4(4):152–153
11. Kansal A, Siddiqui NA, Gautam A (2013) Assessment of heavy metals and their interrelationships with some physicochemical parameters in eco-efficient rivers of Himalayan region. Environ Monit Assess 185(3):2553–2563
12. Karunanidhi D, Aravinthasamy P, Subramani T, Setia R (2021) Effects of COVID-19 pandemic lockdown on microbial and metals contaminations in a part of Thirumanimuthar River, South India: a comparative health hazard perspective416:125909
13. Khanna D, Bhutiani R, Matta G, Singh V, Ishaq FJECJ (2012) Physico-chemical and microbiological status of River Asan in Dehradun Uttarakhand 13(1–2):145–150
14. Khanna D, Singh V, Bhutiani R, Chandra KS, Matta G, Kumar DJECJ (2007) A study of biotic and abiotic factors of Song River at Dehradun 8(3):117–126
15. Khatri N, Tyagi S (2015) Influences of natural and anthropogenic factors on surface and groundwater quality in rural and urban areas. Front Life Sci 8(1):23–39
16. Kulshrestha H, Sharma S (2006) Impact of mass bathing during Ardhkumbh on water quality status of river Ganga37(2):437–440
17. Mondal P, Nandan A, Siddiqui NA (2014) Impact of soak pit on groundwater table 18(1):12–17
18. Mondal P, Nandan A, Siddiqui DN, Yadav BP (2014) Removal of fluoride from water by suitable low cost environmental friendly methods. Environ Pollut Control J 17(6)
19. Nandan A, Siddiqui NA, Singh C, Aeri A, Gwenzi W, Ighalo JO, de Carvalho Nagliate P, Meili L, Singh P, Chaukura N, Rangabhashiyam S (2021) COVID-19 pandemic in Uttarakhand, India: environmental recovery or degradation? J Environ Chem Eng 9(6):106595
20. Nandan A, Yadav BP, Baksi S, Bose DJWSN (2017) Recent scenario of solid waste management in India (66):56–74
21. Nandan A, Siddiqui NA, Singh C, Aeri A, Gwenzi W, Ighalo JO, de Carvalho Nagliate P, Meili L, Singh P, Chaukura N, Rangabhashiyam S (2021) COVID-19 pandemic in Uttarakhand, India: environmental recovery or degradation? J Environ Chem Eng 9(6):106595
22. Nilsson C, Renöfält BM (2008) Linking flow regime and water quality in rivers: a challenge to adaptive catchment management 13(2)

23. Niyobuhungiro RV, Schenck CJ (2021) The dynamics of indiscriminate/illegal dumping of waste in Fisantekraal, Cape Town, South Africa. J Environ Manage 293:112954
24. Padgelwar S, Nandan A, Mishra AK (2021) Plastic waste management and current scenario in India: a review. Int J Environ Anal Chem 101(13):1894–1906
25. Pathak SS, Mishra P (2020) A review of the Ganga river water pollution along major urban centres in the state of Uttar Pradesh, India**7**(3):1202–1210
26. Pierini V I, Mazzeo N, Cazenave M, Semmartin MJR, Conservation and Recycling (2021) Waste generation and pro-environmental behaviors at household level: a citizen science study in Buenos Aires (Argentina) 170:105560
27. Rana D, Joshi N (2023) Assessment of human intervention on ecology of river—a case study. In: International conference on advances and innovations in recycling engineering. Springer, Singapore, pp 51–70
28. Santosh Vani SV, Bhaumik S, Nandan A, Siddiqui NA (2017) Hazardous Waste-Impact on health and Environment for sustainable development in India. World Sci News 70(2):158–172
29. Sasidhar T, Nandan A, Sharma M (2015) Assessment and improvement of safety culture in a sugar industry in India. Res J Eng Technol 6(3):369
30. Seth R, Mohan M, Singh P, Singh R, Dobhal R, Singh KP, Gupta SJAWS (2016) Water quality evaluation of Himalayan rivers of Kumaun region, Uttarakhand, India 6(2):137–147
31. Shrivastava N, Mishra D, Mishra P, Bajpai A (2013) Water quality deterioration of Machna River due to sewage disposal, Betul, Madhya Pradesh, India. J Environ Earth Sci 3(6):1–5
32. Subuddhi SP, Kansal A, Pandey P, Rana D, Ghoshal T. Study of impact of influx waste water generated at vehicle service centre on river water quality characteristics
33. Tiwari A, Dwivedi A, Mayank PJHCR (2016) Time scale changes in the water quality of the Ganga River, India and estimation of suitability for exotic and hardy fishes 7(3):254
34. Tripathi DK, Pandey G, Jain CK (2015) Physicochemical analysis of selected springs water samples of Dehradun City, Uttarakhand, India2(5):99–103
35. Trivedi RK, Goel PK, Karad, India (1984) Chemical and biological methods for pollution
36. Vijayaraman S, Mondal P, Nandan A, Siddiqui NA (2020) Presence of microplastic in water bodies and its impact on human health. In: Advances in air pollution profiling and control. Springer, Singapore, pp 57–65
37. Wan MT, Kuo JN, Pasternak J (2005) Residues of endosulfan and other selected organochlorine pesticides in farm areas of the Lower Fraser Valley, British Columbia, Canada34(4):1186–1193
38. Welfare NCOPMOHF (2019) Population projections for india and states 2011–2036 report of the technical group on population projections New Delhi
39. Yadav A, Siddiqui NA, Nandan A (2020) Impact of water pollution on endemic species, Singapore, Springer Singapore

Sustainable Waste Management of Man-Made Dams and Reservoirs—A Review

Swati Kalra, Omprakash Madguni, and Damini Rana

Abstract Fresh water being an essential component for the survival of life is facing considerable unplanned anthropogenic intervention. To overcome this challenge dams and reservoirs have been made and used for years for the regulation of rivers and ensure that the adequate supply of water can be supplied during dry periods. Reservoirs are artificial formation, usually formed by the construction of a dam across a river by diverting a part of the river flow and storing the water in a reservoir. Upon completion of the dam, the river pools behind the dam and fills the artificially created basin. Seasonal changes of runoff and precipitation feed the reservoir. The stored water can be used for irrigation, drinking water after purification, to produce energy, also for fishing or another livelihood for society. The sustainability of man-made reservoirs lies on the ecosystem services rendered by them. These services can be assessed against the three pillars of sustainability, i.e., ecological/environmental resilience, economically viable and socially accepted and benefitted. This state of art observed that man-made reservoirs gradually acquire all the characteristics of natural water bodies and hence provide ecological services as well as enhancement in the environmental footprint. Study also reveals reservoirs plays a critical role in current situation although the multi-purpose reservoirs should be a priority to lessen the rate of constructions as well as their ecological footprint. Learning suggest that sustainability of reservoirs is dependent on the location, stakeholder participation and its sustainable use.

Keywords Sustainable waste management · Dams and reservoirs · Ecological/environmental resilience · Ecosystem services · Natural water bodies

---

S. Kalra · O. Madguni
IIFM, Bhopal, Madhya Pradesh, India

D. Rana (✉)
Department of Environmental Sciences, Uttarakhand and Central Laboratory, Gurukula Kangri (Deemed to be University), Uttarakhand Pollution Control Board, Kanya Gurukula Campus, Haridwar, India
e-mail: daminiranaofficial@gmail.com

N. A. Siddiqui et al. (eds.), *Advances in Waste Management*, Lecture Notes in Civil Engineering 301, https://doi.org/10.1007/978-981-19-7506-6_11

1 Introduction

A reservoir is a man-made water body or large freshwater body of water constructed for various economic and social developments. This is considered as an alternative to natural water bodies such as lake etc. looking to the large demand for water to meet the domestic, social, Power and Industries. However, the key difference between the lake and reservoir is that reservoirs are artificial and made by humans, while lakes are naturally occurring water bodies. But there are reservoirs constructed in the past for development purposes, which have achieved most of the characteristics of the natural water bodies [48] like lakes and have been conventionally called lakes, for example, the Bhopal Lake. Reservoirs are larger water bodies because they provide a supply of water for when naturally occurring bodies of water, like lakes or rivers, run dry.

Reservoirs can be shaped by a structure or a barrier across a basin, by unearthing the terrestrial or by neighboring a piece of land with barrages and diverting a part of the river flow into the reservoir [9]. The water is deposited in the basin and can be used for multi-purpose work like for irrigation of agricultural field, generation of electricity by hydro-power projects and for domestic & Industrial practices [27, 43]. Artificial reservoirs also play a very current constructions to control unpredicted floods as storm-water supervision. A reservoir is nourished by rainfall, precipitation runoff or from a continuous stream of the nearby waterway. River water loss can arise due to evaporation (especially in arid regions) and depending on the basin bottommost due to filtration (small reservoirs are often lined). Deposits from rivers or surface runoff can lessen the stowage bulk of an artificial basin suggestively [11].

In addition to providing multiple economic benefits to the people, the reservoir can also help to mitigate the negative effects of the water changeability and the extreme environmental actions. Furthermore, such a huge water infrastructure has incurred significant environmental and social costs, spurring demands for natural-based alternatives [41] (Fig. 1).

2 Methodology

Before exploring principles and studies, the researcher tried to figure out the area of interest. Based on the interest, the researcher identified the area and topic, "Sustainability of man-made reservoirs". Based on the theme and objectives, a thorough literature review was planned and for that, various sources were identified, including published articles, grey literature, reports, online media, website etc. literature collected from these sources have been stored/organized and deselecting the information that is useful. Online technical databases which were searched to download relevant literature from elite journals like the Springer journal, Journal of Remote Sensing & GIS Techniques, Inventory of wetlands of Uttarakhand (2012), etc. The collected literature was then thoroughly reviewed to understanding the ecological

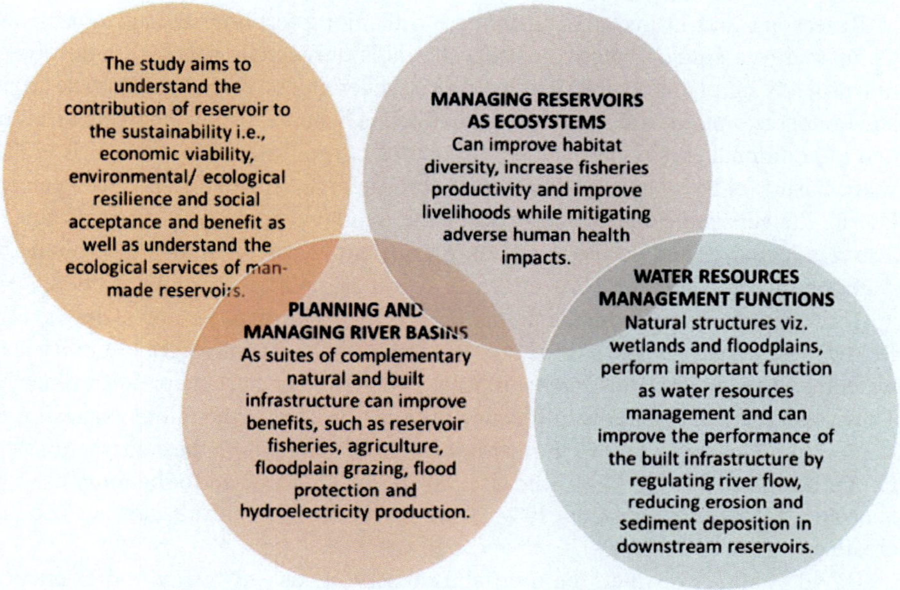

Fig. 1 Benefits of reservoirs

services of man-made reservoirs in India and globally. Based on the reviewed literature, contents have been analyzed and synthesized all the information to fulfil the study objectives.

3 Review and Discussion

After furnishing a general introduction, an attempt has been made in this section to review the literature on the subject. The literature on various aspects of ecological services of manmade reservoirs is quite extensive. In order to find out the major contribution that took place in the ecological services of manmade reservoirs, a brief review of existing literature on different aspects of manmade reservoirs has been made in this section.

If we consider the ancient civilizations, individuals consume making basins for thousands of ages. The eldest acknowledged dam in the world is the Jawa Dam in what is now known as Jordan. It was constructed in nearby 3000 BCE to supply river to usage for irrigation agricultural lands to grow major crops to meet the demand of food [28].

Huge basins alter the ecology of river, and by doing so, they were modify the capability and admission to environment amenities, including overflow directive, fertility of soil along with cultural services such as regeneration [34, 47].

Reservoirs and Dams are constructed with multiple purposes and objectives. Compared to a single objective project that aids only single resolve, multi objective projects are planned for irrigation, power generation, overflow controller, urban and trade, restoration, and fish and flora and fauna benefits, in any amalgamations of two or additional (International Commission of Large Dams (ICOLD) [17]). Folks shape basins for the reason that the volume of water in a stream differs over a period. During the very monsoon period or when the mountain ice is melting, the water in a stream increases and occasionally run-offs its banks. By restrictive the volume of water permissible to endure downriver, reservoirs support controlling of flood.

Throughout dry periods, the water level in a stream can be actual low-slung. Below these situations, the river is unrestricted from the basin so agriculturalists can water their agriculture lands and households and industries can occupation ordinarily. The arid climate and water insufficiency in India directed to the initial expansion of step-wells and water reserve administration methods, including the construction of a Girnar basin in 3000 BC [37]. Likewise, reservoirs are too used for boating, fishing, and other forms of restoration. Few of the reservoirs that form basins are used to create energy.

David et al. [7] distinct, the protruding reservoirs as the "water bodies created by dams built of rubble, earth, stone and masonry work across seasonal streams, as against reservoirs, formed by dams built with precise engineering skill across perennial or long seasonal rivers or streams, using concrete masonry or stone, for power supply, large-scale irrigation or flood control purposes".

3.1 Contribution of the Reservoir to Three Pillars of Sustainability

The notion of Sustainable Growth is distinct as per "development that meets the needs of the present without compromising the ability of the future generations to meet their own needs" through [5].

Furthermore, for the water-based expansion developments the awareness of sustainable progress comprises purpose in addition to arrangement of the stresses for the water through the project area, balanced water use, contribution of the shareholders, equitable growth for anthropological development, accomplishment of the ecological and communal fortification by a functional financial growth, all-inclusive reflection and valuation, active management and establishment of the essential situations for the safety of water-related constructions [30, 44]. Sustainable development in river is essentially proficient by means of the basin water storing bulk in a sustainable method.

3.2 *Importance for Environmental/Ecological Resilience*

3.2.1 Increase the Biodiversity

((BNHS) [2]) found that Nandur-Madhmeshwar is a huge water storing basin, shaped by the structure of a dam at the convergence of the Godavari and the Kadva tributaries. The lake was shaped unplaned in 1907–1913 on the Godavari River as either a resource of irrigation water of agricultural field. Over the decades, the water released from Gangapur and Darana reservoirs is warehoused at Nandur-Madhmeshwar and later released from here through canals for irrigation. Massive amounts of silt and organic matter passed in the earlier 90 years have gathered in the lake, due to which landmasses, surface water ponds and marshlands have been developed. This has caused a decent wetland habitation for birds. The reservoir fills with rainwater runoff between July and September and fascinates several species of migratory birds between September and March [6].

According to [32] lakes, rivers, and many other freshwater ecosystems in India have a diverse biota that represents nearly all taxonomic groups.

Molur et al. [24] discovered that perhaps the lagoon ecological systems established by aquifers of the Western Ghats, a geomorphic provincial capital in southern India that circles the western coast, covers around 136,800 km^2 of total area and are home to an estimated 290 fish species, 77 molluscs species, 171 odonates species, 608 aquatic flora lifeforms, as well as 137 species of amphibians. Also, 53% freshwater fish, 36% aquatic mollusks, while 24% aqueous flora species were accounted in the area.

Likewise, [40] claim that perhaps the Loktak Lake, Manipur, the largest natural water body in North-East India, has the surprising amount of biotic variety. The lake is notable for the floating carpets of rich vegetation, locally known as Phumdi (an distinctive ecology that consists of a diversified type of flora, mud and biological material in various stages of disintegration) and aimed at preserving the endangered Sangai (known as Brow-Antlered Deer in Manipur).

3.3 *Importance of Reservoirs for Economic Viability*

3.3.1 Tourism and Recreational Benefits

Richter et al. [36] in their paper "Restoring Environmental Flows by Modifying Dam Operations" evaluate the reimbursements of barrier "re-operation" that comprises recapture of a variety of fish, shellfish, and other types of fauna inhabitants treasured commercially as well as recreationally, together with the estuarine kind; recrudescence of the overflow storing and aquatic distillation profits that happen when overflows stand permissible to stream into valley of forestry and wetlands; retrieval some

Table 1 Tourism characteristics at the lakes under investigation

Characteristics	Solińskie lake
Tourist traffic	Primarily comprised of multi-day or longer journeys, with half of the respondents visiting this location for the first time
Tourist motivation	Above all, tourist motivation is based on natural environmental assets
Main tourism functions	The main functions of tourism are sightseeing and recreation
Tourism infrastructure	A well-developed, diverse tourism infrastructure with a substantial share of agricultural farms

resemblance of the natural lively equilibrium among waterway loss and sedimentation that forms bodily habitation complication, besides interesting difficulties linked with the geomorphic disproportions [14]; spiritual and cultural purpose of the rivers; and a slew of other valuable goods and services socially.

Duda-Gromada [8] has done his study on the artificial reservoir i.e., Solinskie Lake in Poland originated due to the formation of a man-made reservoir that frequently develops the visitor draw of the area. The educations showed by him consist of a few precise landscapes of travel around the reservoir under discussion. They are presented briefly in Table 1.

3.3.2 Fisheries Production

According to the (FAO), domestic fisheries with a production of 11.47 million metric tonnes in 2015, account for the total global fish production of 12.2%, with marine fisheries accounting for the rest [13]. Because of the difficulties of exactly measuring fish catches (many of which are consumed straight or surplused in discrete unceremonious markets) from a multitude of different sources, this is clearly an undervalue [1].

Fluet-Chouinard [12] discovered in their analyses that river catches are roughly 65% ahead of what national governments publicly record.

The world's biggest domestic capture fisheries are located in developing nations in tropical and subtropical climates, according to [21]. Most are small-scale in nature, but they provide a vital resource for a few of world's weakest and extremely vulnerable populations by contribution towards food security, nutrition as well as providing work. In several of the world's least developed locations, fish is the main source of nutrition protein and income.

In his study, [23] discovered that creating permanent wetlands on reservoir drawdown zones can improve reservoir fisheries. Wetlands like these add towards the richness of habitat by providing refuge, spawning, and nursery grounds for fish that aren't exposed to the drastic differences in water flow that occur throughout the reservoir. Maintaining river connection and safeguarding existing fish habitat (particularly spawning habitat) on flowing down tributaries is often the most significant management strategy to preserving natural reservoir fisheries upstream of reservoirs.

Table 2 Total fish production during 2017–2018 in India

Sectorial contribution of Indian fishers	
Position globally	3rd in the Fisheries 2nd in the Aquaculture
Fisheries to GDP (%) Contribution	0.91
Agriculture. GDP (%) Contribution	5.23
Fish availability (kilograms) in Per capita	9.0
Annually Export earnings in Crores (Rs.)	45,106.89
Sector based employment (million)	14.0

(*Source* (NFDB) [26])

Indian agriculture and fishery regions are an important part of food production, providing nourishment, maintenance support, and meaningful work to over 14 million people, as well as assisting to agricultural exports. The country exhibited constant as well as sustained increases in fish output, since the independence with diversified resources that range from the deep seas to highland lakes and around more than 10% of wildlife species in regards of fish and shellfish species. Overall fish produce in 2017–18 has anticipated to be 12.60 million metric tonnes, with roughly 65% coming from the inland fisheries and almost 50% from cultural fisheries, accounting for around 6.3% of the world fish production (Guledagudda, Reddy et al.) ((NFDB) [26]).

The total production of fish that is 12.60 million metric tons, is accomplished by numerous incomes that are mentioned in Table 2, out of which 3.150 million hectares of the area of the reservoir that is maximum is used for fish production in India as per the report of (NFDB) [26]) (Fig. 2).

3.4 *Importance of Reservoirs for Social Acceptance and Benefits*

3.4.1 Irrigation

In many regions, the capacity to forecast and accomplish rainfall is difficult, with subsequent overflow variability, is an important supplier to food insecurity and poverty. Sadoff et al. [38] mentioned that Water sacristy occurs frequently during periods of adequate water availability, affecting agricultural output and leading to calamities such as floods and droughts. Rainfall fluctuation is an important factor determining economic growth, particularly in largely agrarian cultures [4].

Reservoirs have always been a source of multiple-use water services like irrigation, residential use, recreational use, fisheries, and silt capture, groundwater recharge and flood control as well as flood management [25].

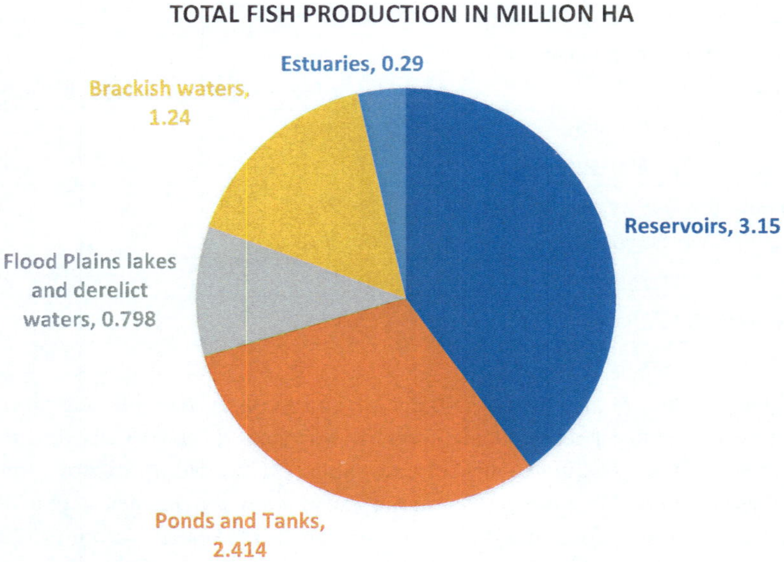

Fig. 2 Fisheries production from different resources in India (*Source* (NFDB) [26])

Tamil Nadu, Andhra Pradesh, and Karnataka considered to possess largest density of the irrigation tanks in India, that accounts for about 60% of India's tank irrigated land (i.e., 0.12 million tanks) [18].

According to a study [31], a critical role is played by tanks in water harvesting, particularly throughout surface-runoff in the monsoon season, which can then be utilized. Conventional tank systems account for roughly 25% of net tank irrigated land in West Bengal, Orissa, Uttar Pradesh, and Bihar.

As per International Commission on Large Dams (International Commission of Large Dams (ICOLD) [17]) and [46], efficient irrigation systems are claimed to convert dry regions suitable for agricultural purposes. Also, reservoirs are served of multiple sources; electricity generation, irrigation supply, reduce the discharge variability for improvement in navigation, and various other usages.

The significance of dams on waterways, discharge as well as irrigated water supply during the twentieth century has been documented [3], with the creation of huge reservoirs considerably boosting global surface water extractions for irrigation. At the beginning of the century, reservoirs provided about 5% of the irrigated areas from the surface water; by end of the twenty-first century, this has raised about 40%. Annual average reservoir irrigation extractions climbed from 18 $km^3 yr^{-1}$ at the turn of the millennium (1901–1920) to 460 $km^3 yr^{-1}$ at the turn of the century (1981–2000) in real terms. The expansion took place mostly on continents having large irrigated areas and multiple dams of irrigation. On the provincial and global scale, irrigated separations and dam management have an influence on the schedule and absolute quantity of water drained into the oceans [33]. In Europe, Africa and Asia, where cumulative surface water discharge into the ocean is 10% lower under some months than in a normalized state, large reservoirs have the most dramatic consequences.

3.4.2 Flood Control

Gupta et al. [16], in their research found India's unique rainfall pattern makes it extremely vulnerable to the floods. Floods threaten 40 million ha of land, or around 1/8th of the country's total land area. Floods claim thousands of lives and millions of dollars in property each year, with losses worsened by fast population increase, uncontrolled urbanization, and unchecked environmental deterioration. To solve the problem, the country has made both structural as well as non-structural efforts. Whereas, non-structural measures like the flood prediction help to increase flood contingency planning by turning society away from the floodwaters, structural measures involve building physical structures [35] such as embankment dams, reservoirs, storm drains, and storage tanks to keep floodwaters from reaching potential damage centers [39].

3.4.3 Ecosystem Services Provided by the Dammed River and Reservoirs

Leemans and De Groot [20] describe ecosystem services as "the benefits individuals gain from ecosystems". These services have proven to be frequently life-saving and improve human well-being. As a result, they are considered to be part of the global commons and are frequently seen as free. Ecosystem services thinking recognizes that nature has value and that this value may be assessed and utilized to inform environmental management decisions [48].

The direct as well as indirect contributions of ecosystems to the human welfare are referred to as "ecosystem services" ((TEEB) [42]). They contribute to organisms' survival and quality of life, either directly or indirectly (Fig. 3).

Ecosystem services can be divided into four categories, according to ((TEEB) [42]):

a. **Provisioning services** include things like food, timber, fiber, fresh water, genetic resources and the medicines that are extracted from ecosystems.
b. **Regulating services** are advantages derived from regulations of ecological systems like climate regulations, water purification and waste management [45], natural hazard regulations [29], pollination, and insect controls.
c. **Habitat services** emphasize, importance of the ecosystems in providing habitat for the migratory species and preserving gene-pool viability.

Fig. 3 Depicting the ecosystem services provided by reservoirs, which are divided into four categories based on the economics of ecosystems and biodiversity

d. **Cultural services** encompass non-materialistic benefits like intellectual development, recreation, aesthetic qualities and spiritual enrichment that individuals gain from ecosystems.

Evaluation of the links between ecosystem services and both manmade and natural infrastructure is critical while building the dams. Researchers defined two categories of water relating ecosystem services from man-made reservoirs in order to do this (Fig. 4).

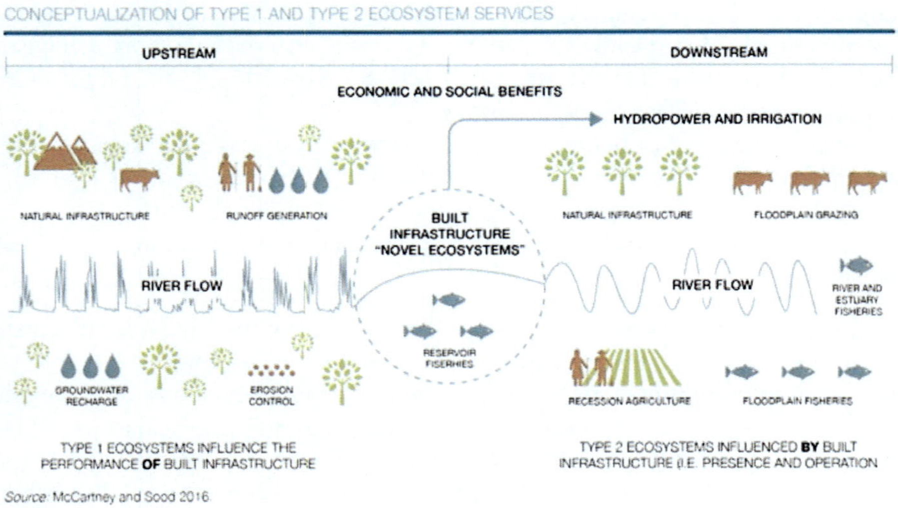

Fig. 4 Upstream and downstream of the reservoir ecosystem services are conceptualized as Type-1 and Type-2 ecosystem services [22]

- **Type 1** ecosystem services affect the constructed infrastructure's technical performance (resilience, vulnerability and reliability) and similarly, its capacity for delivering the advantages of its creation (e.g., irrigation and/or hydropower).
- **Type 2 is** affected by the physical presence of a built infrastructure, as well as changes in water quality, sediment or the nutrient fluxes generated by design and operation of built infrastructure (e.g., floodplain agriculture and/or fisheries); [22].

4 Conclusion

Drought and flood management, hydropower, irrigation, drinking water, navigation, industrial development, recreation, fishing, and other livelihood generation are only few of the uses for reservoirs. According to the literature, after these reservoirs have developed enough, they will take on the features of natural water bodies such as lakes; As a result, reaching the 'mature' reservoir requires the stabilization of all biological and ecological processes, which can take centuries. Even though they are shallow, well mixed, and oxygenated, compact dams are more productive than large, deep ones.

Reservoirs were originally built to provide access to water for irrigation on agricultural land or for domestic use, but later reservoirs also improved local people's livelihoods by allowing them to catch and sell fish in the local market. However, as the population of migrating birds and other animals grows, it contributes to increased diversity and attracts local tourists.

Irrigation, fisheries, flood control, household and industrial water supply, and recreation are just a few of ecosystem goods and services offered by reservoirs. Carbon capture and storage, flood mitigation, groundwater flow, fertilizer clearance, toxics storage, and biodiversity preservation are all valuable services.

The studies carried out clearly demonstrate the reservoir's role in developing tourism and recreation in the area. Tourism began as a social phenomenon, but it has evolved into a substantial economic activity in a short span of time. Tourism is a business for both the individual entrepreneur and the community as an economic activity. Artificial reservoirs are one of the reasons for expanding tourism beyond the economy, which is for aesthetic reasons and to promote the biodiversity of that area, in order to win the economy.

Birds' incredible diversity leads to their importance in ecosystems. The benefits that humans gain from the natural world are referred to as ecosystem services, and birds play an important role in supplying many of these benefits.

A very true line was said by novelist *Jonathan Franzen* that ***"If you could see every bird in the world, you'd see the whole world"*** means that how birds can create the whole ecosystem in one piece of land with limited boundaries but you still see the whole world on that piece of land. Therefore, artificial reservoirs attract most of the migratory birds and other animals too which increase the biological diversity of that area indirectly. The most clearly quantifiable cultural services offered by birds are the leisure activities they support—hunting and birding, for example, both of which support tourism because people go particularly to partake in these activities.

It can be concluded from the review that the creation of reservoirs has been the necessity, but the multi-purpose reservoirs should be the priority to reduce the cost of constructions as well as its footprint. It is also true that the ecological footprint of reservoirs is initially very high, which would be substantiated by way of ecological goods and services. It is also evident that sustainability of the reservoir is dependent on the location, stakeholder participation and sustainable use.

References

1. Allison EH, Mills DJ (2018) Counting the fish eaten rather than the fish caught 115(29):7459–7461
2. (BNHS), B. N. H. S. (2004) Bombay natural history society
3. Biemans H, Haddeland I, Kabat P, Ludwig F, Hutjes R, Heinke J, von Bloh W, Gerten DJWRR (2011) Impact of reservoirs on river discharge and irrigation water supply during the 20th century 47(3)
4. Brown C, Lall U (2006) Water and economic development: the role of variability and a framework for resilience. Natural resources forum, Wiley Online Library
5. Brundtland GH (1987) "Report of the World Commission on environment and development:" our common future, UN
6. Chavan SP, Kulkarni AN, Jadhav PL (2018) Bird and their conservation status at Nandur Madhameshwar bird Sanctuary near Nashik, India (5):109–119
7. David A, Rao N, Ray P (1974) Tank fishery resources of Karnataka, Not Available
8. Duda-Gromada K (2012) Use of river reservoirs for tourism and recreation case study: Solińskie lake in Poland **27**(4)
9. El Gayar A (2020) Impact assessment on water harvesting and valley dams5:266–282
10. FAO, Food and agriculture organization of the united nations
11. FAO (1992) Food and agriculture organization of the united nations
12. Fluet-Chouinard, E., S. Funge-Smith and P. B. J. P. o. t. N. A. o. S. McIntyre (2018). "Global hidden harvest of freshwater fish revealed by household surveys." **115**(29): 7623–7628.
13. Funge-Smith S, Bennett AJF, Fisheries (2019) A fresh look at inland fisheries and their role in food security and livelihoods 20(6):1176–1195
14. Ganguly M, Aynyas R, Nandan A, Mondal P (2018) Hazardous area map: an approach of sustainable urban planning and industrial development—a review. Nat Hazards 91(3):1385–1405
15. Guledagudda S, Reddy SV, Paled MP, Analytical study on indian fisheries sector: trends in trade performance
16. Gupta S, Javed A, Datt D (2003) Economics of flood protection in India. Flood problem and management in South Asia, Springer, 199–210
17. International Commission of Large Dams (ICOLD) (2007) World Register of Dams, Int. Comm
18. Kumar MD, Rao N (2017) The hydro-institutional challenge of managing tanks: a study of tanks in rural Andhra Pradesh, 155–191
19. Large Dams, Paris, France
20. Leemans R, De Groot R (2003) Millennium Ecosystem Assessment: Ecosystems and human well-being: a framework for assessment
21. Lynch AJ, Cooke SJ, Deines AM, Bower SD, Bunnell DB, Cowx IG, Nguyen VM, Nohner J, Phouthavong K, Riley BJER (2016) The social, economic, and environmental importance of inland fish and fisheries 24(2):115–121

22. McCartney M, Sood A (2016) Understanding the link to water and quantifying Ecosystem Services in the Tana Basin, Kenya. Ecosystem Services Partnership Conference
23. Meynell PJH, Dams (2014) Improving reservoir ecology with constructed wetlands 3:78–80
24. Molur S, Smith K, Daniel B, Darwall WJC, UK, Gland SI, I. Z. O. O. Coimbatore (2011) The status and distribution of freshwater biodiversity in the Western Ghats, India
25. Mondal P, Nandan A, Siddiqui NA, Yadav BP (2014) Impact of soak pit on groundwater table. Environ Pollut Control J 18:12–17
26. (NFDB), N. f. d. b (2017–18) National fisheries development board
27. Nandan A, Yadav BP, Baksi S, Bose DJWSN (2017) Assessment of water footprint in paper & pulp industry & its impact on sustainability 64:84–98
28. National Geography (2019) Reservoir
29. Ora A, Nandan A, Kumar A (2017) Hazard identification of chemical mixing plant through hazop study. Int J Adv Res Dev 2(3)
30. Padgelwar S, Nandan A, Mishra AK (2021) Plastic waste management and current scenario in India: a review. Int J Environ Anal Chem 101(13):1894–1906
31. Pant N, Verma R (2010) Tanks in Eastern India: a study in exploration, IWMI
32. Prasad S, Ramachandra T, Ahalya N, Sengupta T, Kumar A, Tiwari A, Vijayan V, Vijayan LJTE (2002) Conservation of wetlands of India-a review 43(1):173–186
33. Priyanka U, Nandan A (2014) Bacterial dynamics in antibiotic wastewater treatment–a review2:001–008
34. Rana D, Joshi N (2023) Assessment of human intervention on ecology of river—a case study. In: International conference on advances and innovations in recycling engineering. Springer, Singapore, pp 51–70
35. Raolji VG, Siddiqui N, Nandan A, Pandya K (2018) A case study on optimization of manual activities through ergonomics interventions13(7):5075–5080
36. Richter BD, Thomas GA (2007) Restoring environmental flows by modifying dam operations 12(1)
37. Rodda JC, Ubertini L (2004) The basis of civilization-water science?, International assn of hydrological sciences
38. Sadoff CW, Hall J, Grey D, Aerts J, Ait-Kadi M, Brown C, Cox A, Dadson S, Garrick D, Kelman J (2015) Securing water, sustaining growth. Report of the GWP/OECD Task Force on Water Security and Sustainable Growth
39. Santana JAD (2020) Terminal-320 de la divisiön territorial de comercializaciön de combustibles de matanzas, Universidad de Matanzas
40. Sharma BK, Sharma S (2009) Faunal diversity of Cladocera (Crustacea: Branchiopoda) of Loktak Lake (a Ramsar site), Manipur (NE India) 106(2):156
41. Siddiqui DN, Nandan A, Sharma M, Srivastava A (2014) Risk management techniques HAZOP and HAZID study. Fire Environ Allied Sci 1(1):5–8
42. (TEEB), T. E. o. E. B (2019) The Economics of Ecosystems & Biodiversity.
43. Subuddhi SP, Kansal A, Pandey P, Rana D, Ghoshal T. Study of impact of influx waste water generated at vehicle service centre on river water quality characteristics
44. Tortajada C (2001) Environmental sustainability of water projects, Institutionen för anläggning och miljö
45. Vijayaraman S, Mondal P, Nandan A, Siddiqui NA (2020) Presence of microplastic in water bodies and its impact on human health. In: Advances in air pollution profiling and control. Springer, Singapore, pp 57–65
46. World Commission on Dams (2000) Dams and development: a new framework for decision-making: The report of the world commission on dams, Earthscan
47. World Commission on Dams (2000)
48. Yadav A, Siddiqui N, Nandan A (2020) Impact of water pollution on endemic species. Advances in water pollution monitoring and control, Springer, pp 49–57

Storage System for Cooking in Hilly Areas: Solar Energy Storage System: Performance Analysis

Geetanjali Raghav, Piyush Kumar, and Mohit Nagpal

Abstract Solar energy storage system finds an application in cooking areas in domestic life especially in the hilly regions of INDIA. For this purpose packed bed system has a better acceptability. Utilization of packed-beds with bed elements with geometrical shapes and sizes and types of rocks and ducts resulted in temperature stratification with an aim to improve the efficiency of solar heater with air as the medium has been proposed by various investigators. However the increase in efficiency also leads to pressure loss; therfoere the system should be robust and designed in such a manner in order to accomodate minimum pressure loss. Pressure losses are generally the function of Reynolds number. (Re., Sphericity (ψ), Void fraction (ϵ), Equivalent diameter (De). So in this work optimization of the solar energy storage system has been performed by considering different shapes and a novel design is proposed. The effect of sphericity over the volumetric heat coefficient and pressure losses is also analysed.

Keywords Energy · Sphericity · Void-Fraction · Pressure Drop

1 Introduction

Solar energy is the most easily accessible or most promising alternative energy option in front of us compared to the fossil fuels. However, the intermittent availability of solar energy leads to an energy gap. So, the question here is why we are talking about all this, and exactly what we are up to. So, the answers are we all are aware of that in Indian villages, electricity is still a major problem and still a large number of Indian villages are deprived of it. The condition is even worse when it comes to hilly areas. Most of the remote villages are disconnected from the mainland in terms of roadways and power lines, considering the fact that there is a huge installation investment and also the returns are minimal. Moreover, India has a rapid growing economy and is

G. Raghav (✉) · P. Kumar · M. Nagpal
Department of Mechanical Engineering, College of Engineering, University of Petroleum and Energy Studies (UPES), Energy Acres, VPO Bidholi, PO Prem Nagar, Dehradun 248007, Uttarakhand, India
e-mail: graghav@ddn.upes.ac.in

N. A. Siddiqui et al. (eds.), *Advances in Waste Management*, Lecture Notes in Civil Engineering 301, https://doi.org/10.1007/978-981-19-7506-6_12

facing a huge energy demand. Also, our focus is now shifting from energy to green energy. The geographical location of the country stands to its benefits for generating solar energy, on referring the data records of the Ministry of renewable energy, India, having a maximum up to 8 to 8.5 DNI (Direct Normal Irradiance) and apart from this India is a tropical country and it receives solar radiation throughout the year.

Proper storage of solar energy could play a vital role in uplifting the living condition of women, especially in hilly regions. Women struggle a lot while cooking during night due to the lack of electricity and proper infrastructure. Moreover, majority of women still use wood and kerosene to cook food, which is not only harmful for their health but also for the nature. With the proper storage and utilization of solar energy, community solar cookers could be very beneficial to them.

Solar energy can be stored during sunny days which is usually more than the energy required to meet the upcoming demand, and at night that charged or stored energy can be used in the form solar air heaters, producing electricity and use it in solar cookers. According to j. s. Saini, Harmeet Singh [1] solar energy can be stored in the form energy related to mechanical, electrical and thermal energy. In all these options thermal energy is considered to be the best and most easily conversable form of energy and it is widely accepted in the market. Solar thermal energy storage system is subdivided into three-parts as shown in Fig. 1.

Sensible heat storage is further divided on the basis of fluid used as liquids and solids. On comparing the liquid and solid storage mediums on the basis of their thermal efficiency liquid sensible heat energy storage is best because they have a high value of sensible heat capacity. Water, salts and many more are being used

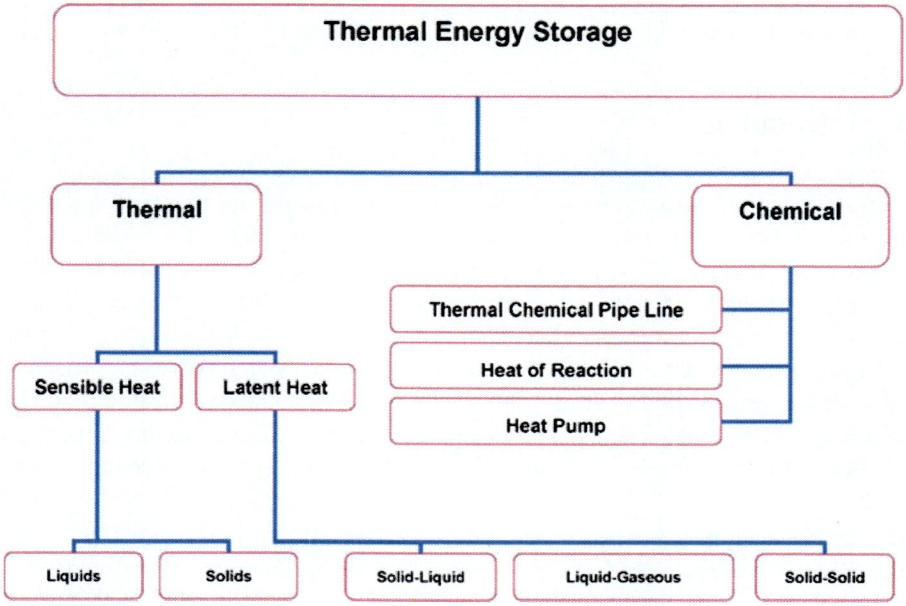

Fig. 1 Different modes of energy storage system

as fluid in the liquid storage system. If comparison is being done on the basis of mechanical characterization, cost, ease of availability, and ease of storage, solid sensible heat storage will be considered a better one.

In the packed bed. solar energy storage system, various things are to be kept in mind in order to optimize the efficiency of the system like Void fraction which is the ratio of free volume to the volume of the container, Sphericity which is a constant value shows how much your shape is close to the shape of sphere. Pressure drop is the drop in pressure from the inlet air pressure, and in order to obtain better efficiency pressure drop should be minimum. Hence the cost of operation of pump should be minimum.

Apart from this, some better container design is also being analysed ranging from cylindrical to triangle and trapezoidal, and different types of materials are also be analysed for better heat storage and efficiency.

2 Literature Review

The heat transfer through the flowing fluid and between fluid and rocks in a packed bed solar energy storage system is firstly given by Schumann [2] in (1930). He studies the heat conduction between fluid and solid and generates the governing partial differential equation for fluid (Eq. 1) and solid (Eq. 2) and shows how their temperature varies with location as well as time.

$$\frac{\partial T_g}{\partial t} + v\frac{\partial T_g}{\partial x} = -\frac{k}{h_g f}(T_g - T_s) \tag{1}$$

$$\frac{\partial T_s}{\partial t} = -\frac{k}{h_s(1-f)}(T_g - T_s) \tag{2}$$

where,

Tg = Temperature of the solid
Ts: fluid tempersture
v: average velocity of the fluid
k: constant heat transfer
hg: heat capacity/fluid
hs: heat capacity/volume of the solid
f: free space/volume

But Schumann model is not fit for practical uses as there are many assumptions being considered before generating these differential equations. Many scientists use Schumann's work as a standard which modified the assumptions like extraction of numerical information from the solution by providing graphs and tabulation given by Klinkenberg [3], ledoux [4], but these equation serve well in hand calculation, and it does not serve well in computational methods using computer calculations.

Due to constraints in the Schumann model, the Mumma and Marvin [5] model/method is being employed that provides an analytical method to compute the temperature stratification, in the bed and the fluid that is based on discretization of system in the number of layers that is explained further in this paper. According to Coutier and Farber [6] gives the finite differential method to compute the temperature of the fluid and solid. We can also consider the farber model as a modification of the Schumann model as he considered the effect of thermal losses, air capacitance, conductance between rocks. Equation (3) for liquid and Eq. (4) for solid are given by Coutier as follows:

$$\frac{\partial T_a}{\partial y} + k_3 \frac{\partial T_a}{\partial z} = T_s - T_a - k_1 (T_a - T_\infty) \tag{3}$$

$$\frac{\partial T_s}{\partial z} = T_a - T_s + k_2 \frac{\partial^2 T_s}{\partial y^2} \tag{4}$$

$$k_1 = \frac{U}{h_v D} \quad k_2 = \frac{h_v k_s}{G^2 c_a^2} \quad k_3 = \frac{\rho_a c_a f}{\rho_s c_s (1 - f)}$$

where, Ta: air temperature; Ts: rock temperature; G: mass air flow rate per unit cross section; Ca: Heat capacity of air; f: void- fraction; D: Rock bed diameter; U: heat loss coefficient of rock bed (W/m^2.c). Here k1 in the equation represents the heat losses to environments and it is cannot be neglected unless the rock bed is super insulated, having initial condition (Eq. 5) and boundary condition (Eq. 6).

$$\begin{aligned} T_a(y, z = 0) &= T_a(y) = T_{a_0} \\ T_s(y, z = 0) &= T_s(y) = T_{s_0} \end{aligned} \tag{5}$$

$$T_a(y, z = 0) = T_a(z) \tag{6}$$

As T(z) is taken constant we can also assume it as a dynamic as it varies with the respect to solar ray and insulation of particular areas. Equations 3 and 4 are further converted into finite differential equations for liquid (Eq. 7) and solid (Eq. 8) as shown below:

$$\frac{T_a(y + \Delta y, z) - T_a(y, z)}{\Delta y} + k_3 \frac{T_a(y, z + \Delta z) - T_a(y, z)}{\Delta z} = T_s - (1 - k_1) T_a + k_1 T_\infty \tag{7}$$

$$\frac{T_s(y, z + \Delta z) - T_s(y, z)}{\Delta z} = T_a - T_s + k_2 \frac{(T_s(y + \Delta y, z) - 2T_s(y, z) + T_s(y - \Delta y, z)}{(\Delta y)^2} \tag{8}$$

$$h_v = 700 (\frac{G}{D_e})^{0.76} \tag{9}$$

Hasnain [7] performed an analysis on the use of various strage materials for low and high temperature storage application. The pebble beds or rocks are generally used as the storage material because of their low cost. According to Rajesh maithani [8] optimization of temperature stratification is accompanied by loss in pressure drop (increase). On the basis of J.s. Saini [8] temperature stratification is a function of sphericity, void- fraction and equivalent diameter and operating parameters such as temperature rise parameter and insolation. As, according to duffle and Beckman [9] the overall performance of the system is totally dependent on temperature stratification or temperature distribution that depends on the operating parameter and he suggested that the recommended size of these particles is 0.01–0.03 m. Firstly, experiments being done by using pebble bed as the material in packed bed results in high pressure drop in the bed resulting in large consumption of power from the fan suggested by Aw Mawire and Pearson [10]. According to Coutier and ard reported that consumption of fan energy should be compared to the maximum energy available. According to Torab [11] it is reported to have. maximize variation of availability of energy to total pumping power which increases with the increment in size of element. So, the large material is being used as a storage material to minimize the pressure drop but according to Sagara and Nakahara [12] the biggest disadvantage of a large size material is having lesser heat transfer due to the lesser surface area. of contact but in contrast of his minimizing the pressure drops, it is considered as efficient. This pressure drop is depending upon the void fraction which depends upon the path created for the flow channel by the packed bed materials suggested by Schmidt and Willmot [13]. Works were performed on bed for accurate prediction of pumping power as suggested by Chandra and Willits [14].

3 Methodology

Considering Mumma and Marvin model as a base for our theoretical calculation. We limit our calculation between Nusselt Numbers, Pressure Drop, Sphericity, Void Fractions and Many more as described in the further mentioned equations:

$$Nu = 0.437 \times \mathrm{Re}^{0.75} \times \Psi^{3.35} \times \varepsilon^{-(1.62)} \times e^{\frac{29.03 \times \log_{10} \psi^2}{D_e^2}} \tag{10}$$

$$h_v = \frac{k_a \times 0.437 \times \mathrm{Re}^{0.75} \times \Psi^{3.35} \times \varepsilon^{-(1.62)} \times e^{\frac{29.03 \times \log_{10} \psi^2}{D_e^2}}}{D_e^2} \tag{11}$$

And the terms Reynolds number(Re), effective diameter(De), void-fraction(Є), Sphericity(ψ) are being found via:

$$D_e = \left(\frac{6V_e}{\pi}\right)^{1/3} \tag{12}$$

$$\varepsilon = \frac{V_b - V_s}{V_b} \tag{13}$$

$$\Psi = \frac{A_s}{A_e} \tag{14}$$

$$\mathrm{Re} = \frac{G \times D_e}{\mu_a} \tag{15}$$

where; Ae: surface area of material element, m^2;
As: surface area of a sphere having volume equal to material element, m^2;
Vb: volume of packed bed, m3; Ve: volume of material element, m^3;
Vs: Total volume of storage material packed in the bed, m^3.
Pressure drop in the bed has been determined using the friction factor reported by Singh, Ranjit (2006) [15]

$$f = 4.466 \times \mathrm{Re}^{-0.2} \times \Psi^{0.696} \times \varepsilon^{-2.945} \times e^{11.85(\log \Psi)^2} \tag{16}$$

$$\Delta p = L \times G^2 \times 4.466 \times \mathrm{Re}^{-0.2} \times \Psi^{0.696} \times \varepsilon^{-2.945} \times e^{\frac{11.85(\log \Psi)^2}{\rho_a D_e}} \tag{17}$$

Singh et al. [15] reported an extensive research on five different types of material sin shape and size and find out the curves between various parameters taking the Mumma and Marvin model as a base.

(1) T-joint Masonry tile brick ($\psi = 0.55$)
(2) Standard Masonry tile bricks ($\psi = 0.63$)
(3) Standard Masonry bricks ($\psi = 0.72$)
(4) Concrete Cubes ($\psi = 0.80$)
(5) Concrete Spheres ($\psi = 1.00$)

After the evaluated curves between Nusselt number and void fraction they found that the packing of the current phase of solid with different shapes.In the case of other shapes usually it gets separated from the material surface as sharp corners are present there. The smoothness of the surface increases with decrement in sphericity.So, our aim is to design two different shapes of the material. We modified them by giving the edge fillet of 2 cm, and changing the material composition also will look into the effect of its change on void fraction, friction factor, Volumetric Heat coefficient and Nusselt Number or Pressure Drop. Detailed Data like area, volume is being found out with the help of CATIA V5 software as shown below (Figs. 2, 3 and 4).

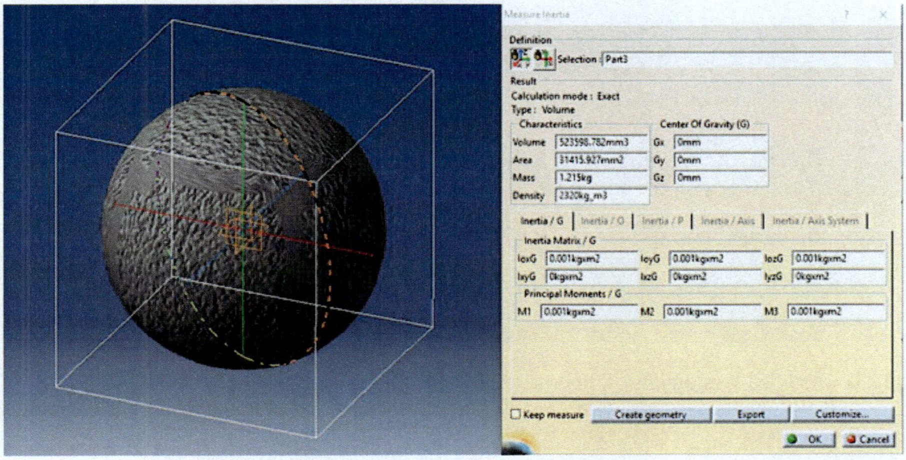

Fig. 2 Sphere of radius 50 mm and having sphericity of 1.00

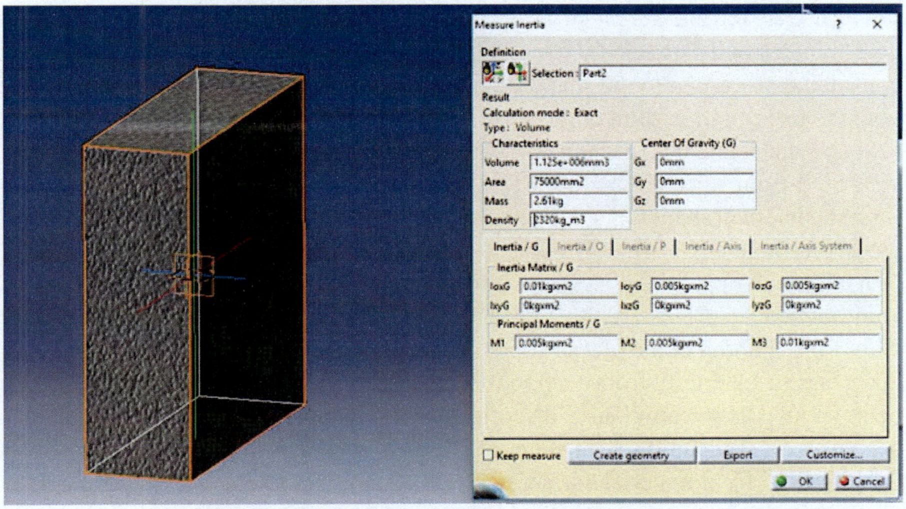

Fig. 3 J.s. Saini Considered shape apart from sphere of sphericity 0.70

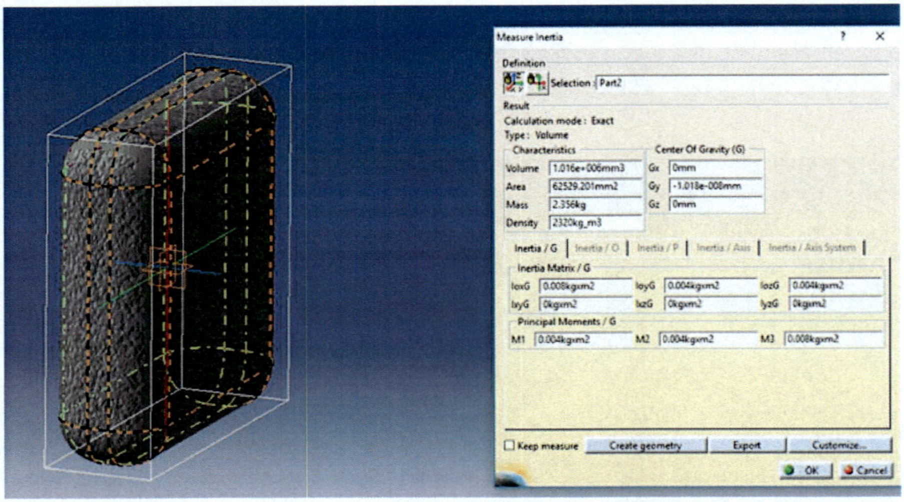

Fig. 4 Modified version of material with 2 cm edge fillet having sphericity of 0.80

4 Conclusions

So, Figs. 5 and 6 depict the changes in flow patterns and the area of contact available for heat transfer are meant to be the reason behind the change of coefficient of heat transfer and Nusselt number. If packing of spherical and others shapes are considered according to the shapes, it appears to be that the fluid remains in contact with the maximum portion of the surfaces of elements of spherical shape in contrast to shapes like cube with a sharp corner at the edge due to which the flowing fluid get detatched from surface. Therefore, lesser surface area is available for heat transfer for non-spherical shapes, but our approach toward removing these sharp corners with the appreciable edge fillet of radius 2 cm that provides us a better heat transfer as proved/shown in graph, as it is much closer towards the spherical shape at a constant void fraction.

In Figs. 7 and 8 in the plot of friction factor and Reynolds no. we noted that with decreases in sphericity from 1.00 to 0.70, the friction factor decreases the reason of being simple as we previously discussed in first paragraph that the area of contact decreases as compared to the spherical shaped ones due to the detachment of the flowing fluid from material surfaces as well as strong corners. So, less contact less friction. It causes decreases in friction factor which results to decrement in pressure drop also.

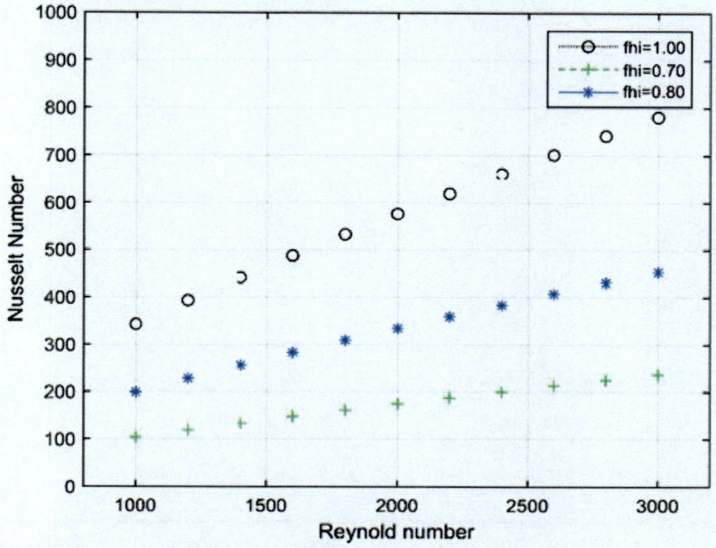

Fig. 5 Plot between Nusselt No. and Reynold No. at constant Void fraction of 0.40

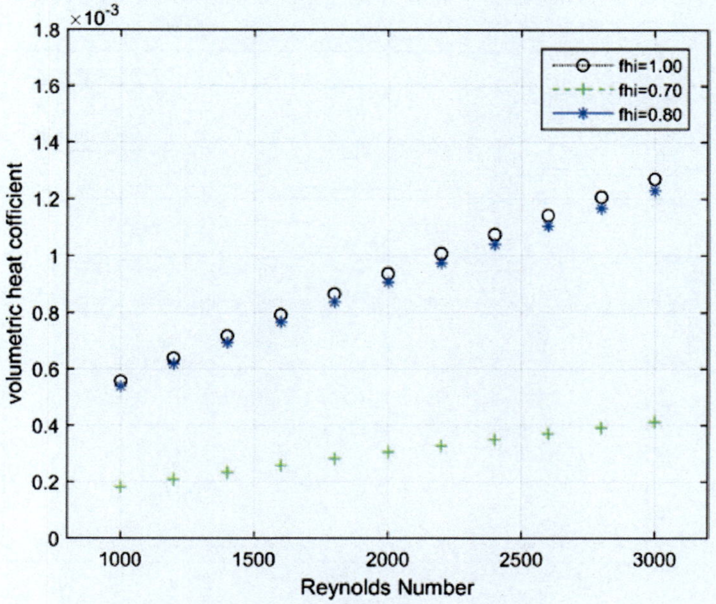

Fig. 6 Plot between hv and Reynolds no at constant void fraction of 0.40

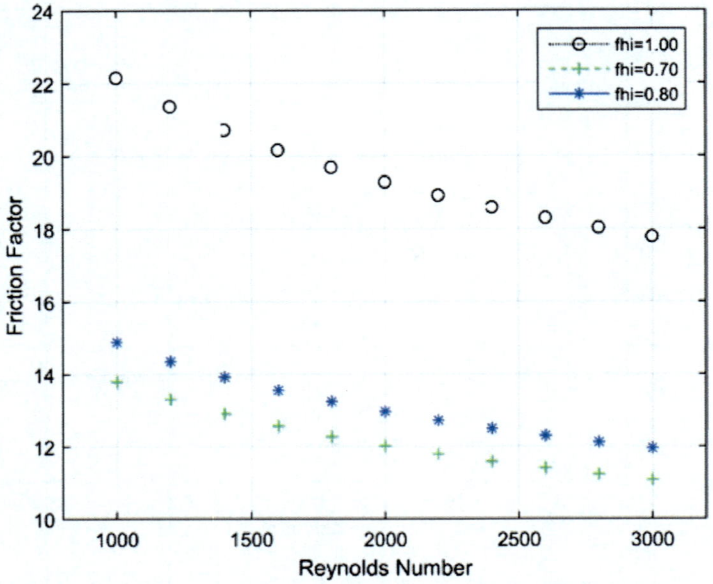

Fig. 7 Plot between Friction factor and Reynolds no. at constant void fraction of 0.40

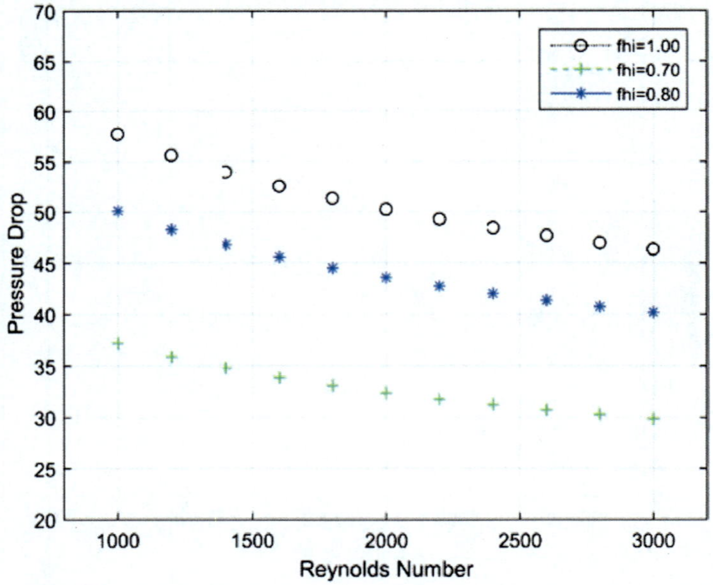

Fig. 8 Plot between Pressure drop and Reynolds no. at constant void fraction of 0.40

References

1. Harmeet S, Saini RP, Saini JS (2010)A review on packed bed solar energy storage systems. Renew Sustain Energy Rev 14(3):1059–1069
2. Schumann TEW (1929) Heat transfer: liquid flowing through a porous prism. Frankline lnst 208:405–416
3. Klinkenberg A (1948) Numerical evaluation of equations describing transient heat and mass transfer in packed solids. Ind Eng Chem, 40
4. Ledoux E (1948) Dynamic cooling at absorbent beds. Ind Eng Chem, p 40
5. Mumma SA, Marvin WC (1976) A method of simulating the performance of a pebble bed thermal energy storage and recovery systems. ASME Paper 76- HT-73, ASME/AIChE national heat transfer conference, St.Louis
6. Coutier JP, Faber EA (1982) Two applications of a numerical approach of heat transfer process within rock beds. Solar Energy 6(29):451–462
7. Hasnain SM (1998) Review on sustainable thermal energy storage technologies. Part 1. Heat storage materials and techniques. Energy Convers Manage J 11(39):1127–38
8. Rajesh M (2013) investigation of effect of stratification on the thermal performance of packed bed solar air heater. Int J Energy Sci (IJES) 3(4)
9. Duffie JA, Beckman WA (1991) Solar engineering of thermal processes, 2nd Ed.; Wiley
10. Mawire A, McPherson M (2009) Simulated performance of storage materials for pebble bed thermal energy storage (TES) systems. Appl Energy 86(7–8):1246–1252
11. Torab H, Beasley DE (1987) Optimization of packed bed thermal energy storage unit. Sol Energy Eng 109:170–175
12. Sagara K, Nakahara N (1928) Thermal performance and pressure drop of packed beds with large storage materials. Sol Energy 47(3):157–163
13. Scmidt FW, Willmott AJ (1981) Thermal energy storage and regeneration. Hemisphere Publishing Corporation, Washington, p 342
14. Chandra P, Willits DH (1981) Pressure drop and heat transfer characteristics of air rock bed thermal storage systems. Sol Energy 6(27):547–553
15. Singh R, Saini RP, Saini JS (2006) Nusselt number and friction factor correlations for packed bed solar energy storage system having large sized elements of different shapes. Sol Energy 80:760–771

A Fuzzy Logic Index-Based Assessment of the Water Quality of River Ganga at the Place of Its Origin (Devprayag) and the Downstream Town of Hrishikesh

D. M. Arjun, N. Pagadeeswar, Tabassum-Abbasi, Abhishek Nandan, Tasneem Abbasi, and S. A. Abbasi

Abstract The river Ganga (the Ganges) is not only one of the largest rivers in India, but it is also revered by the Indians as its most pious and auspicious. The pollution of Ganga in the course of its flow through the Gangetic plain has been exponentially increasing in recent decades, turning the river into one of the most polluted of the world's major rivers. The present work describes a 3-year assessment of the Ganga water quality at the town of its origin—Devprayag—and a town Km downstream: Hrishikesh. The findings indicate.

Keywords Ganga · Fuzzy index · Water quality · Pollution control

1 Introduction

Two glaciers in the Himalayas, at Alkapuri and Gangotri, form the origin of the rivers Alaknanda and Bhagirathi. These rivers meet at Devprayag, between latitude 30° North and 78° East (Fig. 1), giving rise to the river Ganga which then traverses 2525 km, receiving a right bank tributary Yamuna and several left bank tributaries, on way to Bangladesh. There it is joined by Brahmaputra and Meghana before emptying into the Bay of Bengal.

While Ganga is unquestionably a mighty river in terms of its length, catchment, discharge, and its impact on millions of people inhabiting its massive drainage basin, it has all even greater attributes. In the Indian tradition, the water of the Ganga is believed to possess the virtues of de-polluting and de-contaminating all those objects that come in its touch. Even more significantly, Ganga is believed to have extraordinary powers of healing and cleansing even the souls of those who bathe in its water. Scientific studies have been conducted from time to time [1, 2] to identify the factors

D. M. Arjun · N. Pagadeeswar · Tabassum-Abbasi (✉) · A. Nandan
School of Engineering, University of Petroleum and Energy Studies, Bidholi, Dehradun 248 007, India
e-mail: tab.abbasi@gmail.com

T. Abbasi · S. A. Abbasi
Centre for Pollution Control and Environmental Engineering, Pondicherry University, Kalapet, Puducherry 605014, India

N. A. Siddiqui et al. (eds.), *Advances in Waste Management*, Lecture Notes in Civil Engineering 301, https://doi.org/10.1007/978-981-19-7506-6_13

Fig. 1 The origin of river Ganga at Devprayag

that confer extraordinary pathogen-killing ability on Ganga. These studies have led to interesting pointers which are continuing to be explored. But even if the extra virtues of Ganga are kept aside for a separate scientific study, the river's contribution in supporting the livelihood of millions of people habituating the river's >1 million Km^2drainage basin marks Ganga out as one of the world's major natural resource and life-line. For these and other reasons, a great deal of attention has been paid to monitoring the water quality of the river Ganga. India's Central Pollution Control Board has set up a network of 57 water quality monitoring stations, encompassing the entire river basin lying in India [3]. Each station monitors 9 core parameters that most strongly influence the Ganga water quality. The latest report on this monitoring covers the period up to 2013.

The present work updates the information on Ganga water quality at the location of the river's origin at Devprayag (Fig. 1) and at Hrishikesh (Fig. 2), which is the first major pilgrimage centre downstream. A three-year time period has been spanned, covering 2016–2019. A distinguishing feature of the work is that a fuzzy logic-based water quality index has been developed with which the Ganga water quality can be described in terms of a two-digit numeric, thus making the portability as well as navigation of information on Ganga water quality not only much easier, but also effective and evocative. How indices serve this purpose, and how the use of fuzzy logic in indice development enables us in overcoming the weakness of the conventional indices—which are based on crisp arithmetic—has been detailed elsewhere [4–6].

Fig. 2 Ganga at Hrishikesh

2 Methodology

The broad methodology has been outlined in Fig. 3.

2.1 *Identifying Sampling Stations and Acquiring Water Quality Data*

From the network sampling stations set up by CPCB [3], two were identified for this study—Devprayag and Hrishikesh. Data for the years 2016–2019 was acquired from these stations for the index-based analysis reported in this chapter.

2.2 *Development of a Fuzzy Water Quality Index*

This involved the following steps:

(1) Selecting the key parameters that are representative of the overall water quality
(2) Generating membership functions based on water quality standards
(3) Obtaining the fuzzy relationship matrix for different levels of water quality
(4) Performing fuzzy composition and aggregation.

Selecting the key parameters that are representative of the overall water quality

As earlier stated, the Ganga water quality data was obtained from the network of monitoring station, already set up by the Central Pollution Control Board (CPCB). CPCB has identified ten parameters that best represent the Ganga water quality:

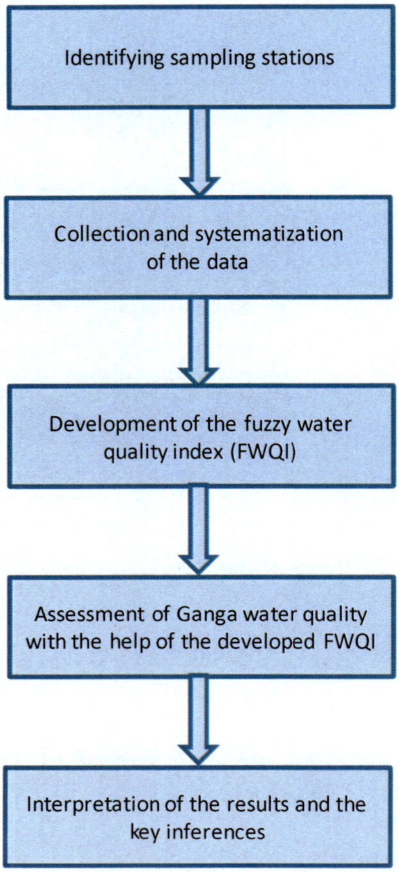

Fig. 3 The broad methodology

total dissolved solids (TDS), hardness, pH, biological oxygen demand (BOD), total coliform, dissolved oxygen (DO), alkalinity, chlorides, calcium, and magnesium.

Formulation of membership functions

Membership functions are used to characterize the fuzziness of discrete or continuous elements of a fuzzy set in the form of a graph [7]. In this work, three-level trapezoidal membership functions were constructed for the development of the water quality index.

To each water quality parameter, the following levels of membership were assigned: "low", "medium" and "high". The levels denoted as "high" indicate the best water quality, followed by lesser levels of "medium" and "low". The degree of membership in the categories of low, medium and high was set based on water quality standards.

These formed the fuzzy input variable/set. Straight-line *trapezoidal* membership functions (MFs) were chosen in this step due to their simplicity and adequacy for this purpose. T*rapezoidal* functions are defined by their flat top. This feature of the

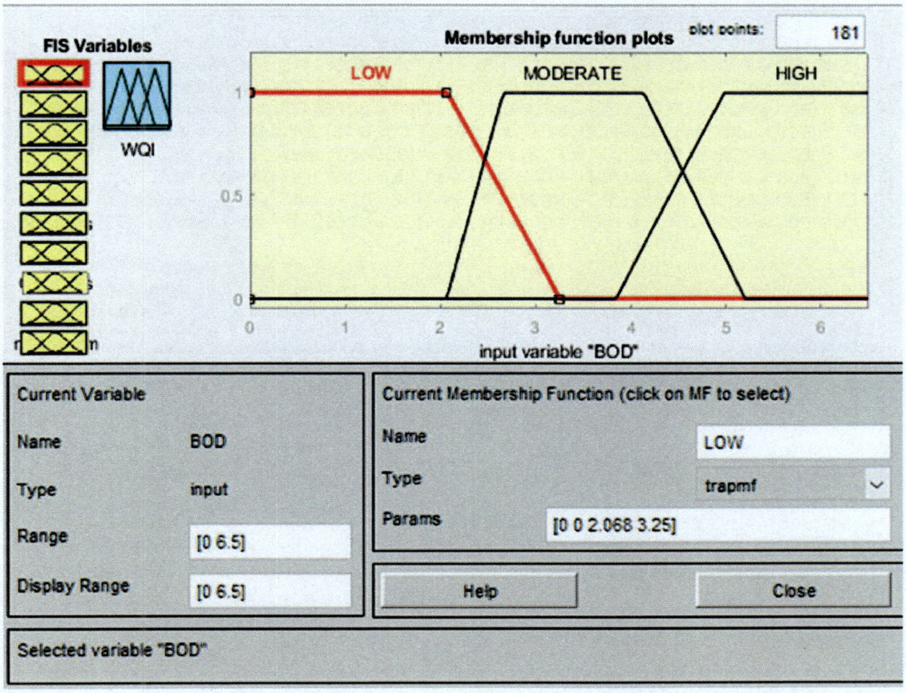

Fig. 4 Membership function for BOD (*Source* MATLAB)

function allows one to have a range of input values that could possess a membership value of 1. The limits of the membership functions were set on the basis of expert judgement and various water quality standards.

The membership function constructed for water quality parameter BOD is shown in Fig. 4.

Setting up an inference engine

The Mamdani fuzzy inference system was used in this step owing to its simplicity and wide applicability [8, 9]. It synthesizes a set of linguistic control rules obtained from human experience into fuzzy operations. The indicators described in the preceding section were fuzzified by the MFs to form the antecedents. The antecedents were mapped to the consequent using fuzzy AND operator and if-then-else rules. Figure 5 shows the rule editor used in MATLAB. All the rules were given an equal weightage of 1.

Defuzzification to arrive at a crisp score

The output of the aggregation process is usually a fuzzy set that has a range of values. In order to convert it back into a single crisp value, defuzzification is done. Among many defuzzification techniques, the centroid method (also known as centre of area or centre of gravity method), which returns the centre of area under the aggregated

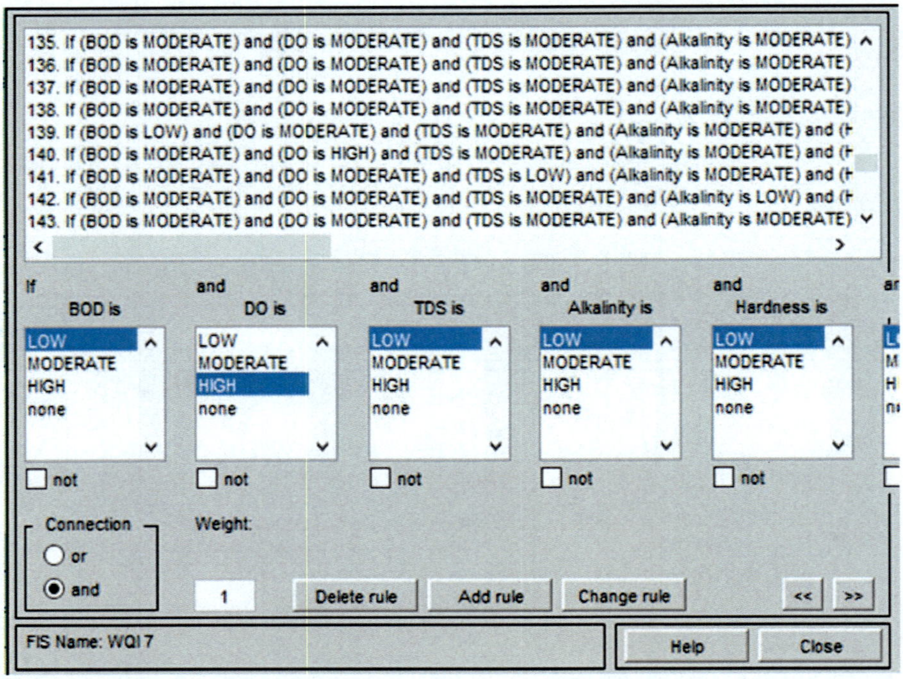

Fig. 5 IF-THEN-ELSE rules as implemented in the MATLAB rule editor toolbox

output curve, is the most prevalent and physically appealing of all the defuzzification methods [7]. Thus, this method of defuzzification was chosen for this study. The final output of this step was the FI-based water quality index.

The FWQI scale

The FWQI scale was set as 0–100; the lesser the value poorer the quality. It was seen that, based on the limits prescribed by the Central Pollution Control Board, India, a water source with FWQI value 70 and above will be fit for drinking and symbolizes "excellent" water quality. A water source with FWQI values between <70 and ≥60 would be considered "good" as it could be used for drinking after subjecting it to turbidity removal and disinfection. A water source of FWQI value <60 and ≥50 will be adequate for bathing, laundry, and similar non-potable uses. A water source with FWQI lesser than 50 will be considered "poor" and will need elaborate treatment.

3 Results and Discussion

The water quality data at Devprayag and its downstream town Hrishikesh has been summarized in Tables 1 and 2. Whereas total dissolved solids, which indicate the gross presence of anions and cations besides dissolved organics, are in the range

Table 1 Summary of water quality at Devprayag, 2017–2019

Parameter	2017		2018		2019	
	Range	Average and SD	Range	Average and SD	Range	Average and SD
TDS, mg/l	64–67	66 ± 1	62–68	66 ± 3	68–74	71 ± 3
Hardness, mg/l	72–76	74 ± 2	66–74	71 ± 3	64–72	68 ± 3
Chlorides, mg/l	5–6	6 ± 1	5–6	6 ± 1	5–6	5 ± 1
Alkalinity mg/l	56–64	61 ± 3	56–64	61 ± 3	60–62	61 ± 1
Calcium, mg/l	38–50	42 ± 5	38–42	40 ± 2	38–46	41 ± 3
Magnesium, mg/l	26–34	32 ± 4	28–32	31 ± 2	20–32	27 ± 6
DO, mg/l	9–9	9 ± 0	9–10	10 ± 0	10–10	10 ± 0
BOD, mg/l	<1	0 ± 0	<1	0 ± 0	<1	0 ± 0
pH	8–8	8 ± 0	8–8	8 ± 0	8–8	8 ± 0
Coliforms (Total)	0–0	0 ± 0	0–0	0 ± 0	<2	0 ± 0

62–74 mg/L at Devprayag during the 2017–2019 span, they appear in a much higher concentration range at Hrishikesh: 62–148 mg/L; the upper concentration being almost double at Hrishikesh. The upper limit of harness at Hrishikesh—126 mg/L—is also nearly twice as much as at Devprayag. But the most alarming difference is in terms of total coliforms, which indicates faecal and other pathogenic contamination. From their near-absence at Devprayag, the Ganga water acquires substantial concentrations of total coliforms by the time it reaches Hrishikesh. These findings indicate that the river Ganga starts getting polluted as soon as it leaves its place of origin (Devprayag).

The pattern of change in water quality across months and years are shown in terms of the calculated FWQI index values for Devprayag in Fig. 6 and for Hrishikesh in Fig. 7. The water quality at Devprayag doesn't seem to vary drastically from month to month and its values have remained in the range 63–75. In contrast, the water quality at Hrishikesh, as reflected in its FWQI values, is seen to fluctuate widely, falling to its lowest value of 46 in one of the months and rising to its highest value of 76 in another. On several occasions, it falls below the worst value, 63, recorded at Devprayag. These findings confirm quantitatively that which was indicated qualitatively by the water quality data summarized in Tables 1 and 2.

At both Devprayag and Hrishikesh, the water quality in 2019 is generally better than seen in 2018 and 2017. This is likely due to the National Mission of Cleaning Ganga (NMCG) or the *Namami Gange* initiatives taken by the Government of India to reduce Ganga's pollution. However, the impact is only marginal and does not appear strong enough to prevent the gross pollution that occurs downstream. It is

Table 2 Summary of water quality at Hrishikesh, 2017–2019

Parameter	2017		2018		2019	
	Range	Average and SD	Range	Average and SD	Range	Average and SD
TDS, mg/l	78–148	96 ± 22	74–126	90 ± 14	62–114	82 ± 13
Hardness, mg/l	70–126	84 ± 14	70–82	77 ± 4	60–78	70 ± 5
Chlorides, mg/l	5–8	6 ± 1	5–5	5 ± 0	4–8	5 ± 1
Alkalinity mg/l	66–76	72 ± 4	64–76	71 ± 4	58–73	64 ± 5
Calcium, mg/l	40–88	48 ± 13	38–46	42 ± 2	36–46	42 ± 3
Magnesium, mg/l	28–40	36 ± 3	30–40	35 ± 3	20–34	28 ± 5
DO, mg/l	10–11	10 ± 0	9–10	10 ± 0	10–12	10 ± 1
BOD, mg/l	<1	0 ± 0	<1	0 ± 0	<1	0 ± 0
pH	7–8	7 ± 0	7–8	8 ± 0	7–8	8 ± 0
Coliforms (Total)	22–80	49 ± 20	26–70	43 ± 15	40–60	46 ± 7

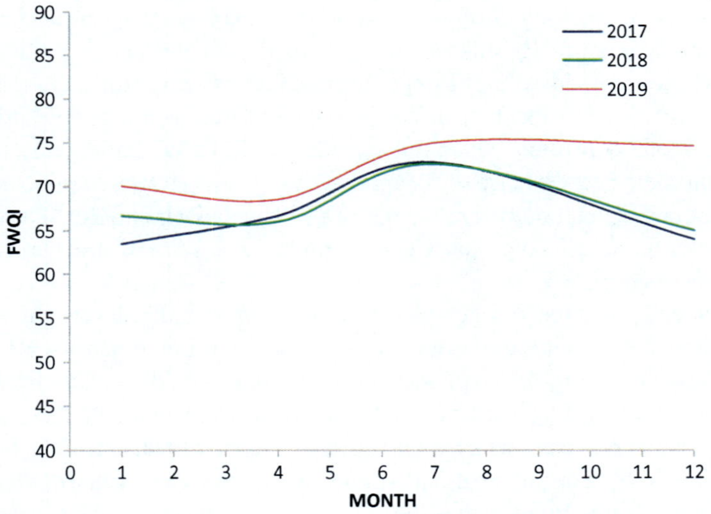

Fig. 6 Fuzzy water quality index (FWQI) across different months and years at Devprayag

imperative that inexpensive wastewater treatment systems with low carbon footprints are deployed to control the pollutants coming in [10], and economics of pollution control [11–20]. More extensive analysis of the constituents is also indicated [21] along with period risk assessment [22, 23].

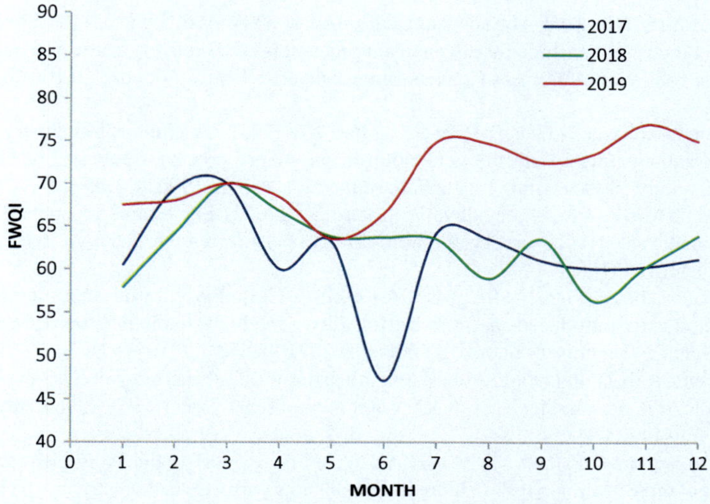

Fig. 7 Fuzzy water quality index (FWQI) across different months and years at Hrishikesh

References

1. Koshy J (2016) Healing powers of Ganga, https://www.thehindu.com/news/national/Labs-study-healing-powers-of-Ganga/article60566726.ece
2. Hashmi K (2020) Healing waters: India's Ganges River cleanses itself during virus shutdown-The Mainichi. https://mainichi.jp/english/articles/20200501/p2a/00m/0fe/023000c
3. CPCB (Central Pollution Control Board, 2013): Pollution Assessment-River Ganga, cpcb.nic.in
4. Abbasi SA (2002) Water quality indices, state of the art report, National Institute of Hydrology, scientific contribution no incoh/sar-25/2002, Roorkee: INCOH, 73–200
5. Tabassum-Abbasi, Abbasi T, Tauseef SM, Abbasi SA (2018) An accident vulnerability index based on fuzzy logic. In Siddiqui NA, Tauseef SM, Abbasi SA, Rangwala AS (eds) Advances in fire and process safety, Springer Transactions in Civil and Environmental Engineering, Chapter 24, pp 291–300. https://doi.org/10.1007/978-981-10-7281-9_24
6. Abbasi T, Khan FI, Tabassum-Abbasi, Abbasi SA (2020) Assessment of the ambient air quality of a highly industrialized suburb of a typical Indian city part 1: assessment of quality. In: Siddiqui N, Tauseef S, Abbasi S, Khan F (eds) Advances in air pollution profiling and control, Springer Transactions in Civil and Environmental Engineering. Springer, Singapore. Chapter 5, pp 67–102. https://doi.org/10.1007/978-981-15-0954-4_5
7. Ross TJ (2004) Fuzzy logic with engineering applications. Wiley, New York
8. Icaga Y (2007) Fuzzy evaluation of water quality classification. Ecol Ind 7(3):710–718. https://doi.org/10.1016/j.ecolind.2006.08.002
9. Mathworks: (2022). https://www.mathworks.com/help/fuzzy/types-of-fuzzy-inference-systems.html
10. Tabassum-Abbasi Patnaik P, Tauseef SM, Abbasi SA (2022) SHEFROL® bioreactor enhances the ability of pistia (*Pistia stratiotes*) in the phytoremediation of greywater. Int J Environ Anal Chem 102(7):1502–1511. https://doi.org/10.1080/03067319.2020.1738420
11. Hussain N, Abbasi SA, (2018) Efficacy of the vermicomposts of different organic wastes as "clean" fertilizers: state-of-the-art. Sustainability 10:1–63. https://doi.org/10.3390/su10041205
12. Hussain N, Abbasi T, Abbasi SA (2016) Vermicomposting transforms allelopathic parthenium into a benign organic fertilizer. J Environ Manage 180c:180–189

13. Tauseef SM, Tabassum-Abbasi, Patnaik P, Abbasi T, Abbasi SA (2021) A novel high-rate vermicomposting machine for downstream processing of biorefinery waste and other forms of organic solid waste. Biomass Convers Biorefinery 1–10. https://doi.org/10.1007/s13399-021-01522-w
14. Tabassum-Abbasi, Patnaik P, Rahi R, Abbasi SA (2022) A circular biorefinery-integrating wastewater treatment with the generation of an energy precursor and an organic fertilizer. Sustainability (Switzerland) 14(9):5714. https://doi.org/10.3390/su14095714
15. Abbasi T, Mishra PK, Shreevidhya R, Abbasi SA (2022) Exploration of organic wastes and problematic weeds as absorbents for the removal of oil and grease from greywater. Ind J Environ Protect 42(5):631–636
16. Tabassum-Abbasi, Patnaik P, Abbasi SA (2022) A Closed loop wastewater biorefinery based on the recently introduced shefrol reactor and two freely available macrophytes. Clean–Soil, Air, Water, 1–12. https://doi.org/10.1002/clen.202100287
17. Tabassum-Abbasi, Patnaik P, Abbasi SA, Abbasi T (2022) Feasibility of a novel (SHEFROL) technology in pre-treating eatery wastewater at pilot scale. Water 14(22):3789. https://doi.org/10.3390/w14223789
18. Tabassum-Abbasi, Patnaik P, Abbasi SA (2022) Process design and assessment of the performance of three macrophytes in a biorefinery polishing partly treated sewage in novel SHEFROL bioreactors. Processes 10(11):2350. https://doi.org/10.3390/pr10112350
19. Tabassum-Abbasi, Abbasi T, Abbasi SA (2019) Microbial fuel cells as source of clean energy—potential and pitfalls. Nat Environ Pollut Technol 18(3):789–797
20. Abbasi SA, Tabassum-Abbasi, Patnaik P (2021) Biomimetically generated nanoparticles in boosting the performance of microbial fuel cells. Nat Environ Pollut Technol 20(5):1881–1886. https://doi.org/10.46488/NEPT.2021.v20i05.003
21. Abbasi SA (1989) Atomic absorption spectrometric and spectrophotometric trace analysis of uranium in environmental samples with Np-methoxyphenyl-2-furylacrylohydroxamic acid and 4-(2-pyridylazo) resorcinol. Int J Environ Analyt Chem 36(3):163–172
22. Tauseef SM, Rashtchian D, Abbasi SA (2011) CFD-based simulation of dense gas dispersion in presence of obstacles. J Loss Prevent Process Ind 24:371–376
23. Tabassum-Abbasi, Abbasi T, Abbasi SA (2018) Role of HAZOP in assessing risk of accidents in chemical process industries: capability and lacunae. Int J Eng Sci Math 7(1):482–487

Eco-sensitive Campus Development Planning: A Case Study

Tabassum-Abbasi, K. B. Chari, Tasneem Abbasi, and S. A. Abbasi

Abstract Over the years geographical information system (GIS) tools have become less and less expensive as also ever easier to use. This chapter presents a case study in which GIS has been used with essential ground truth validation to set-up 'bench-marks' of land-use, flora, fauna, and topography of a typical campus—that of Pondicherry University—with which development planning can be done in an eco-sensitive manner. The effort also led to the identification of rainwater harvesting sites.

Keywords Eco-centric campus · GIS · Thematic mapping · Environmental management · Rainwater harvesting

1 Introduction

Even as governmental directives as well as citizen's desires aim at eco-sensitive development of campuses and neighbourhoods, it is being rarely done in practice. 'Green development' is generally equated with planting of trees in available spaces. There, too, recourse is generally taken of a few easy-to-establish social forestry species. While such efforts are certainly better than no efforts, their positive impact is only marginal.

But much more informed, systematic, and effective development planning of any area, big or small, can be done if the area is mapped and understood in its totality. With such an approach all avenues of resource conservation can be strengthened while

Tabassum-Abbasi (✉)
School of Engineering, University of Petroleum and Energy Studies, Dehradun 248007, India
e-mail: Tab.abbasi@gmail.com

K. B. Chari
GIS Labs, Hyderabad 500076, India
e-mail: dr.kbchari@gislabs.in

T. Abbasi · S. A. Abbasi
Centre for Pollution Control and Environmental Engineering, Pondicherry University, Kalapet, Puducherry 605014, India

N. A. Siddiqui et al. (eds.), *Advances in Waste Management*, Lecture Notes in Civil Engineering 301, https://doi.org/10.1007/978-981-19-7506-6_14

maximising resource use as well. In other words, proper and total integration can be done of the needs of development with the minimization of its adverse impacts. Achieving it is no longer as expensive or complicated as it used to be before the advent of geographical information system (GIS) tools. Over the years GIS tools have become more and more inexpensive as well as navigable. Even more importantly, such tools enable a continuous review and upgradation of the state of the region to which the tools have been applied.

This chapter presents a case study of deployment of GIS in the integrated eco-centric planning of a specific campus—Pondicherry University.

2 Methodology

2.1 *The Study Area*

The campus has a sloping terrain, flat and even in most part. From its western peak situated 36 m above the mean sea level (MSL), the terrain slopes down to its eastern boundary which is about 7 m above the MSL (Fig. 1). The high tide mark of the sea is about 500 m from that campus edge.

A very distinctive landscape feature is provided by a ravine which runs through the campus from west to east, creating a split down the middle. The ravine is characterized by sharp undulations, especially in its southern flank. The unevenness of the ravine catchment separates that part of the campus terrain from an otherwise predominantly flat landscape (Fig. 2).

2.2 *Ground Survey*

In order to facilitate field survey, the study area was notionally divided into a grid of 18 units (Fig. 3). Each unit was then extensively surveyed for (a) physical features such as roads, buildings, playgrounds, and water tanks, (b) vegetation consisting of flora and monocultures, (c) land use and land cover pattern, (d) topography and elevation contours. For each unit, four or more maps were generated to accommodate all essential details that were surveyed. The composite maps were then prepared by the McHarg's overlay technique [1].

Detailed topography of the area was surveyed with the specific objective of identifying potential sites for rainwater harvesting and groundwater recharge. The elevation contours were surveyed using indirect method of contouring. In this method, contour levels were taken at 13 selected sites. Each site was marked on ground with a grid of one meter interval and the readings were taken diagonally using the dumpy level equipment. The emphasis of the rest of the survey was to identify and quantify sites for eco-restoration.

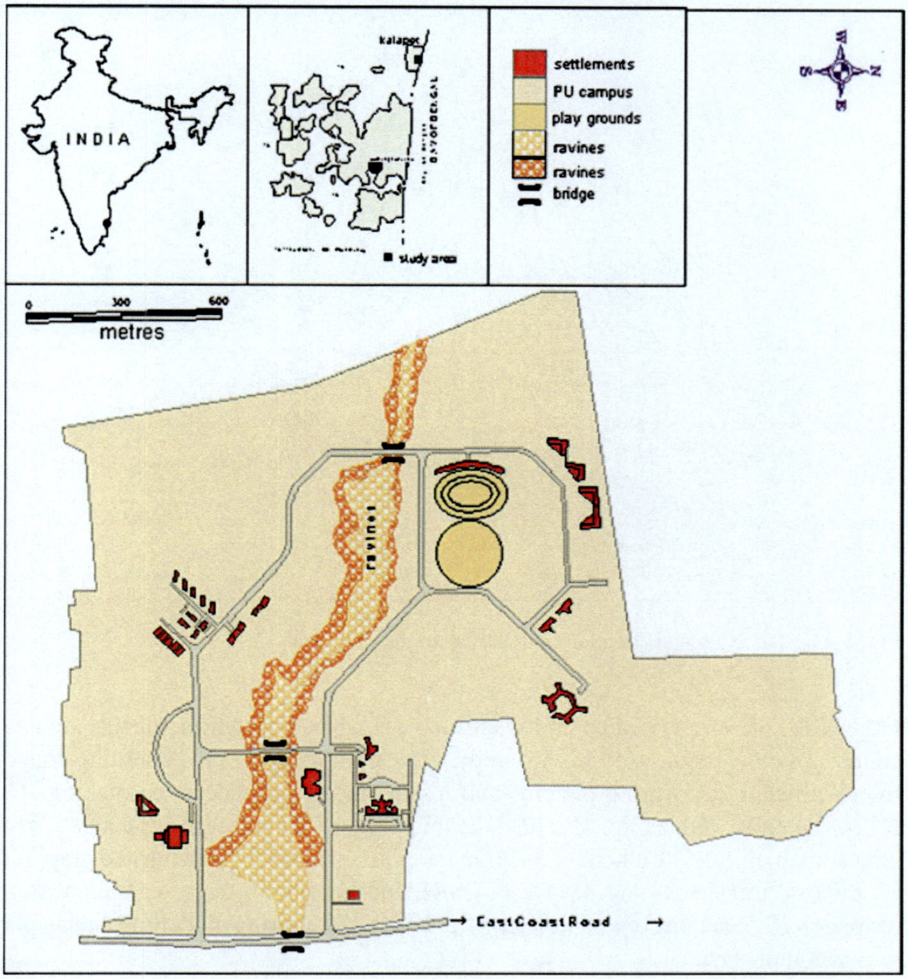

Fig. 1 Location map of the study area (Pondicherry University campus)

2.3 *Digital Mapping*

The paper maps and ground-truth survey maps were scanned with HP Scanjet 6300 colour scanner. All the subsequent processing for GIS was done with MapInfo Professional Vr 5.5 and its accessory Map Viewer developed by Chari and Abbasi [2]. Terrain contours were generated using Vertical Mapper Vr 2.5, an add on tool of MapInfo Professional. The entire process of GIS development for the study area is depicted as flow chart in Fig. 4.

The GIS software, MapInfo Professional, currently marketed as Mapinfo Pro [https://www.precisely.com/pr]. It is one of the leading desk-top mapping solutions capable of sophisticated and detailed data analysis from geographic perspective.

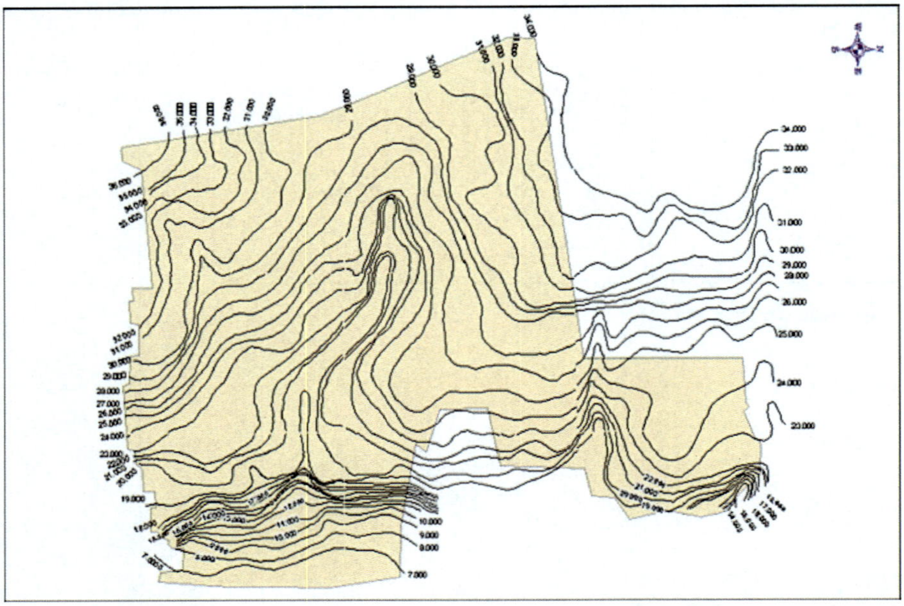

Fig. 2 One metre interval elevation contour map of the study area

Other GIS software of similar ability are also available, including some that are open source. In the version used in this study, also, high quality maps could be created that facilitated discovering patterns and trends for effective decision making. The following tools were explored: Thematic mapping, • Raster image support, • Attach data to map objects, • Linked views, • Geographic searches, • Charting and graphing, • Universal translates for Auto Desk, ESRI, and Intergraph formats. With Vertical Mapper Vr 2.5, and in conjunction with field work, contouring modelling and display was accomplished.

3 The Salient Findings

Figures 5 and 6 provide illustrative examples of the 72 thematic maps and 18 overlayed maps generated by us. Of each grid unit the four thematic maps comprised of (a) physical features, (b) bushes, (c) other plants, and (d) monocultures. This is illustrated in Fig. 5. An overlay of these thematic maps for each grid unit was then generated, as illustrated by Fig. 6. The icons used for the purpose are as indexed in Fig. 7. To save space only one set of maps have been included in this chapter as an example. Full information is available in Abbasi and Abbasi [3].

Thirteen sites (Fig. 8) were identified for rainwater harvesting, based on the micro-level contour surveys. The results of the survey of Site 1 have been given in Fig. 9

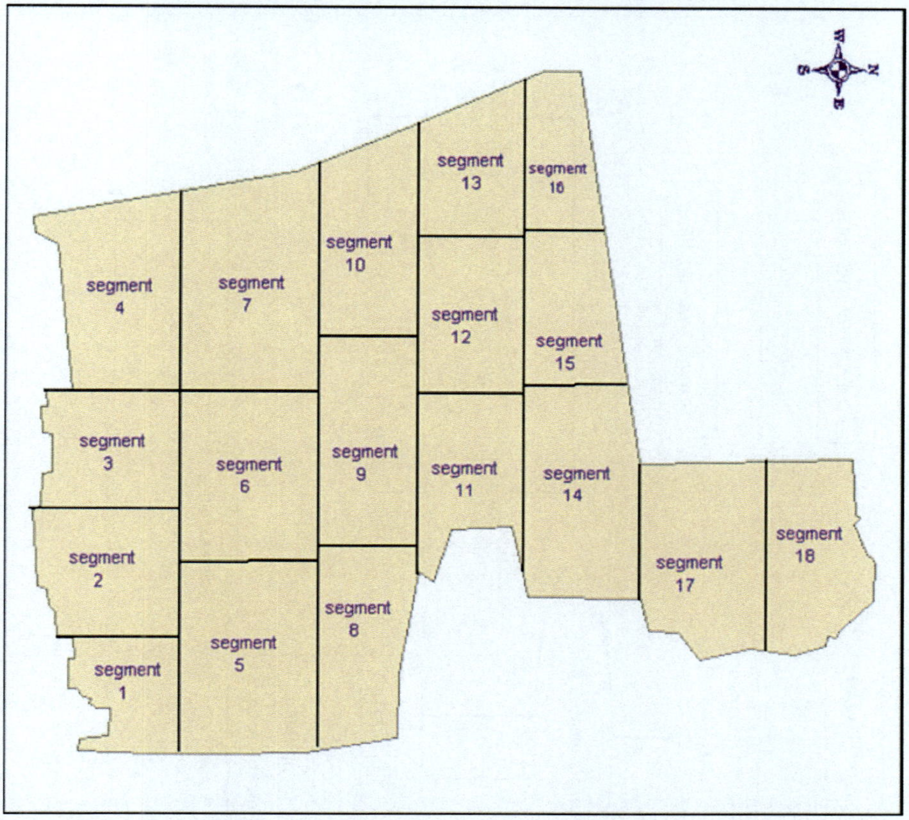

Fig. 3 Notionally divided grids for further detailing of the study area

as an illustrative example. The details of other surveys are available in Abbasi and Abbasi [3].

3.1 Topography

Prior to the taking over of the land, in 1986, for bringing up Pondicherry University, it sported natural vegetation interspersed with planted cashew trees. The green cover kept the soil bound which could resist detachment even in times of intense rainfall.

But once the campus development began, soil was moved about for constructing roads and buildings. At many parts the hold of the natural vegetation over the topsoil was weakened. The campus became increasingly prone to soil erosion. This was exacerbated by the sloping terrain of which the maximum height is 29 m above its base height at the MSL, the campus became increasingly endangered by sheet as well as gully erosion. Now all rains are accompanied by substantial loss of top soil by sheet erosion as well as soil loss due to the run-off forming gullies. There

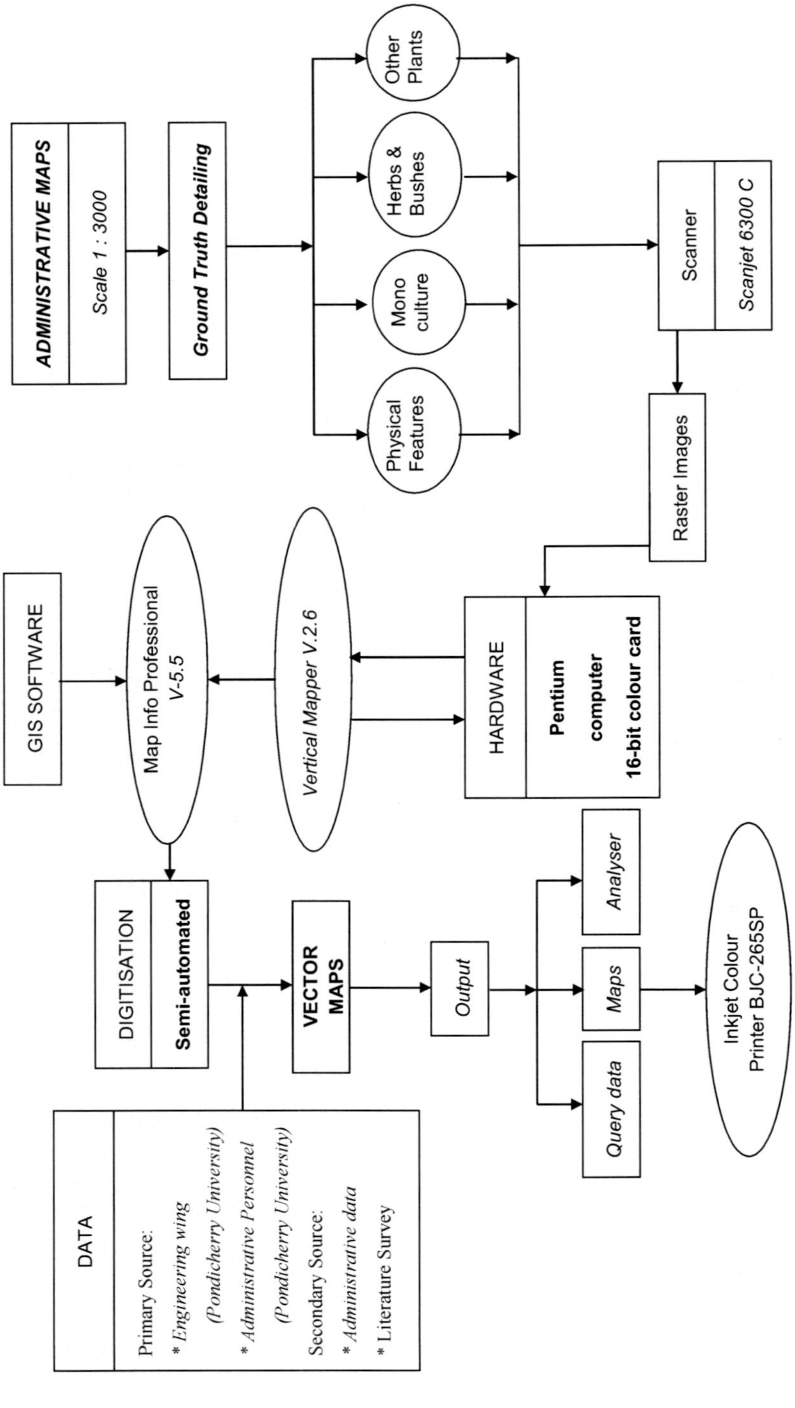

Fig. 4 Flow chart showing the process of GIS development for the present study

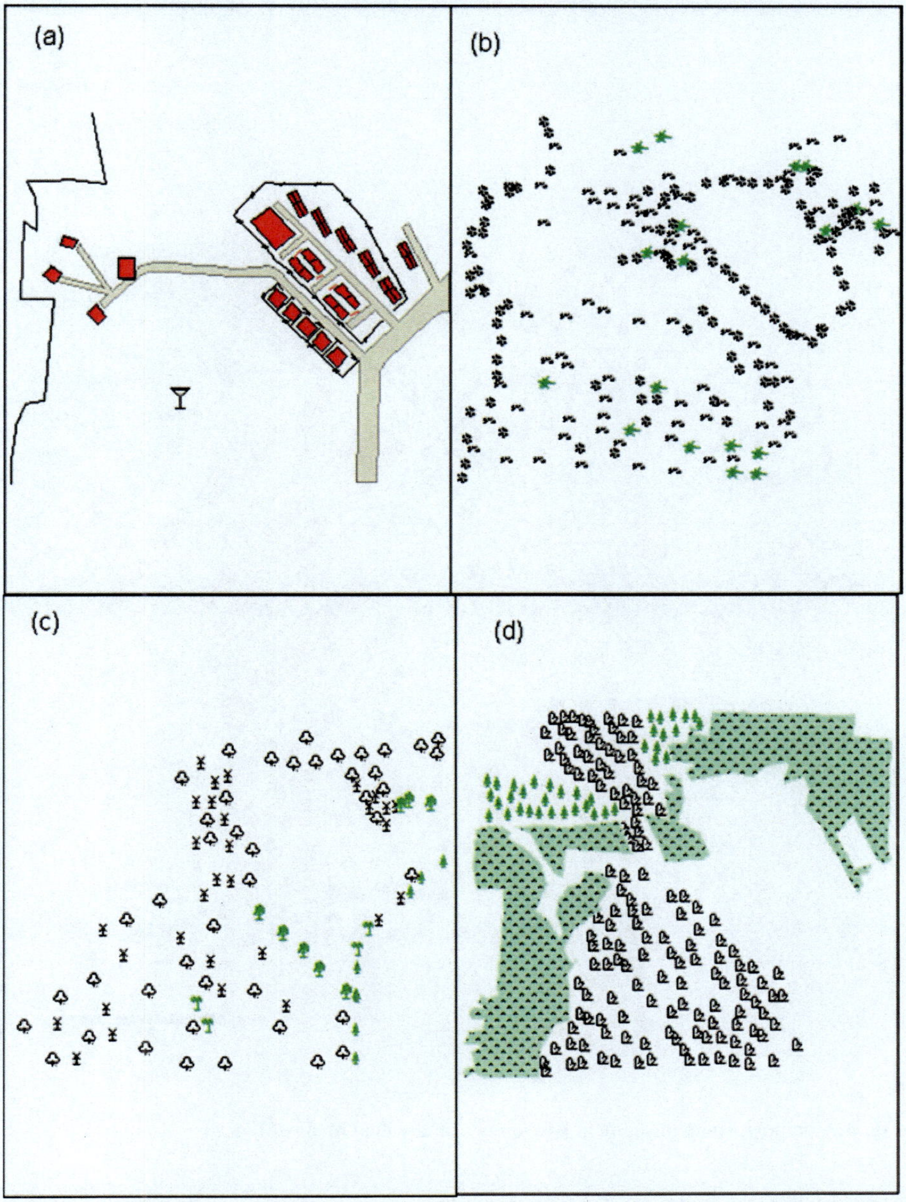

Fig. 5 Feature wise map of one of the segments: **a** physical features, **b** bushes, **c** other plants, d monoculture

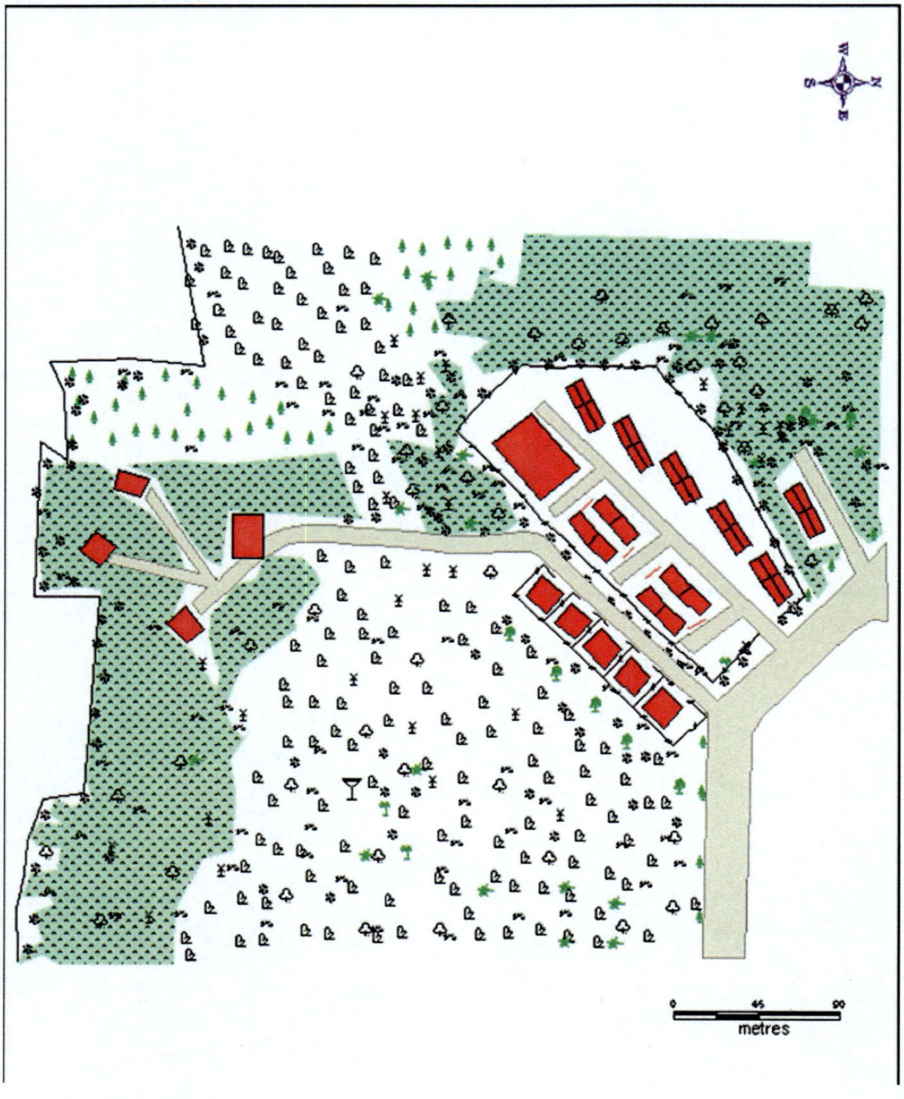

Fig. 6 Composite map of segment 3 (overlay of individual maps of Fig. 5)

is very little scope for the soil that is eroded from one place to collect at another. As a result the topsoil keeps washing off with the rains into the sea. During spells of heavy rains the entire campus suffers soil erosion to a more or less extent. The loss of finer soil particles contributes towards loss of water-holding capacity of the campus terrain. In recent years rainfall is tending to occur at high intensity over a few days than a much more lower intensity rainfall that used to occur earlier and which used to be evenly distributed over larger number of rainy days. Heavy downpour and resulting copious run-off prevents percolation and re-settlement of eroded soil.

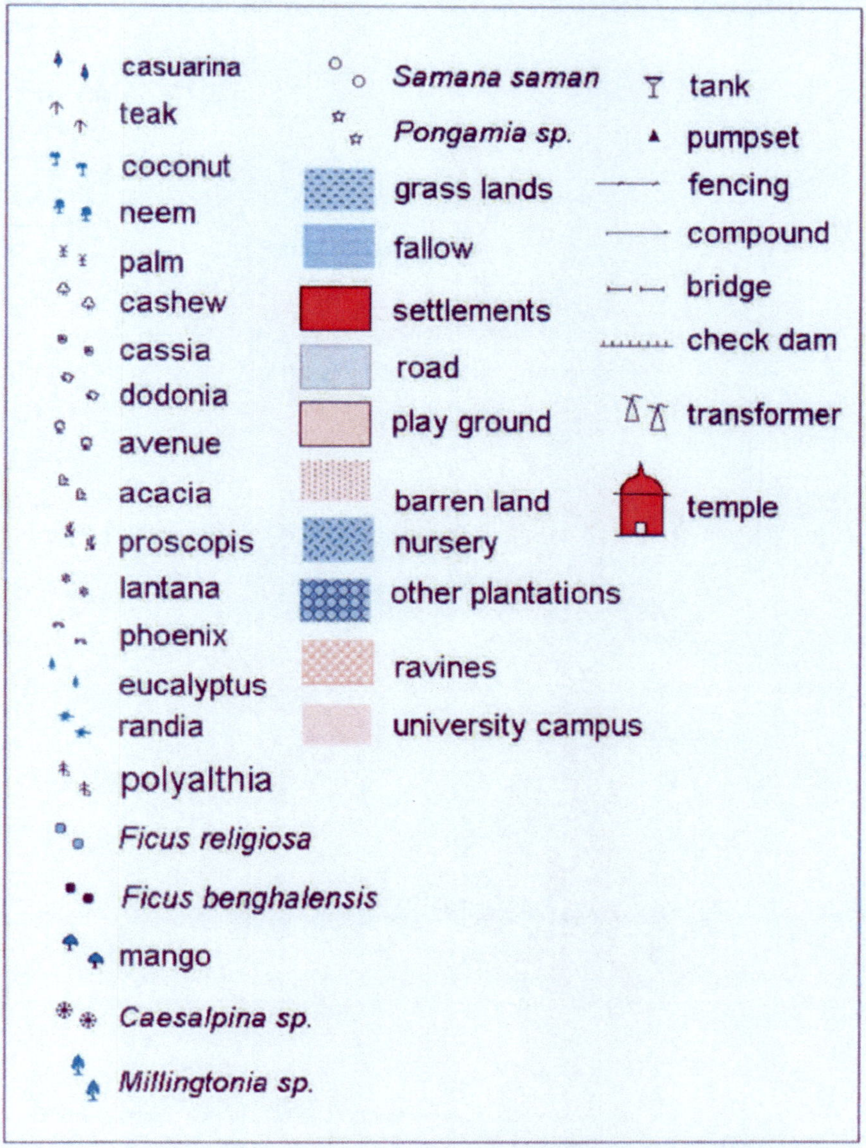

Fig. 7 Legend of icons used in Figs. 5 and 6

Many of the depressions that had existed and which enabled runoff water to collect and get percolated earlier, are now covered by roads or buildings. To prevent than from getting water-logged the campus engineers have constructed several cobbled channels to hasten the run-off. The vegetation around buildings and roadsides is periodically removed on the request of the campus dwellers who believe it will

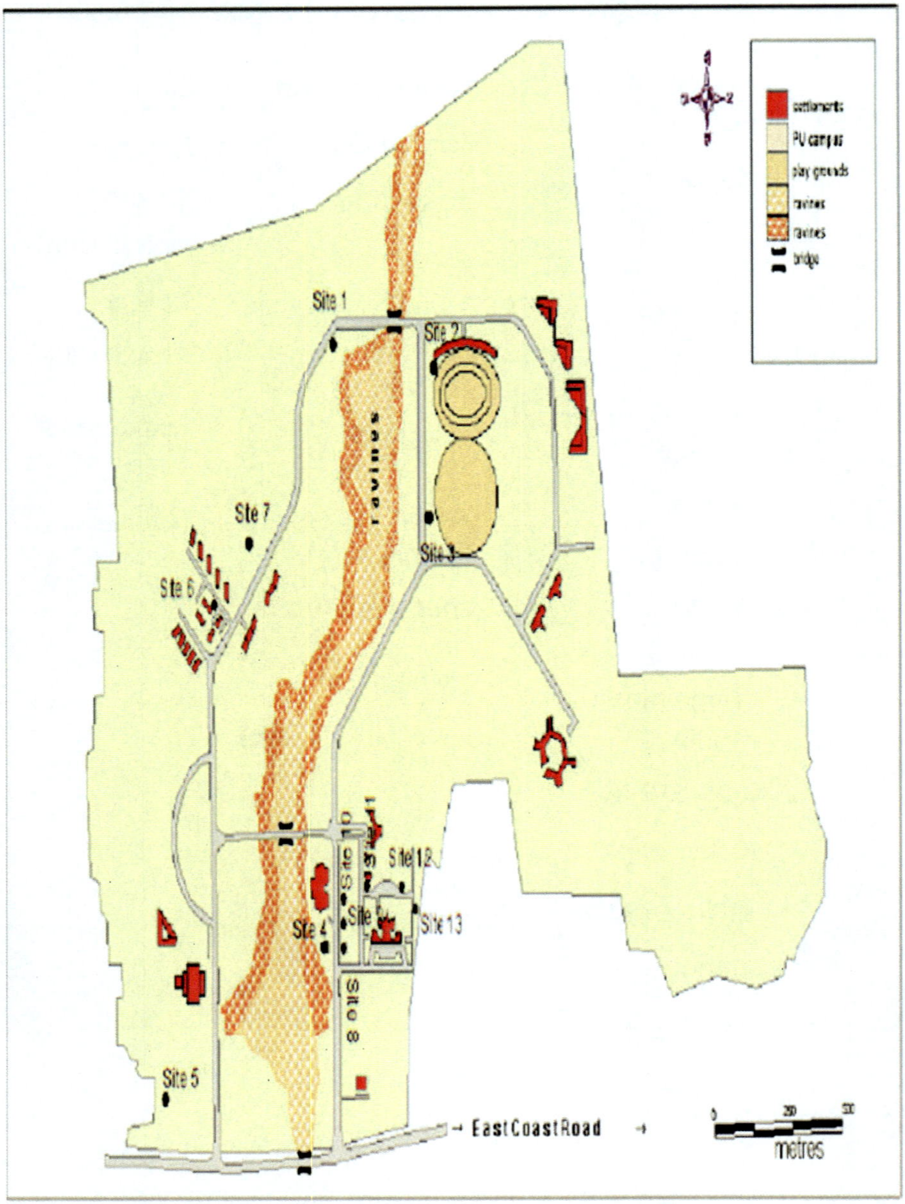

Fig. 8 The locations of the 13 sites identified for the micro-level contour survey

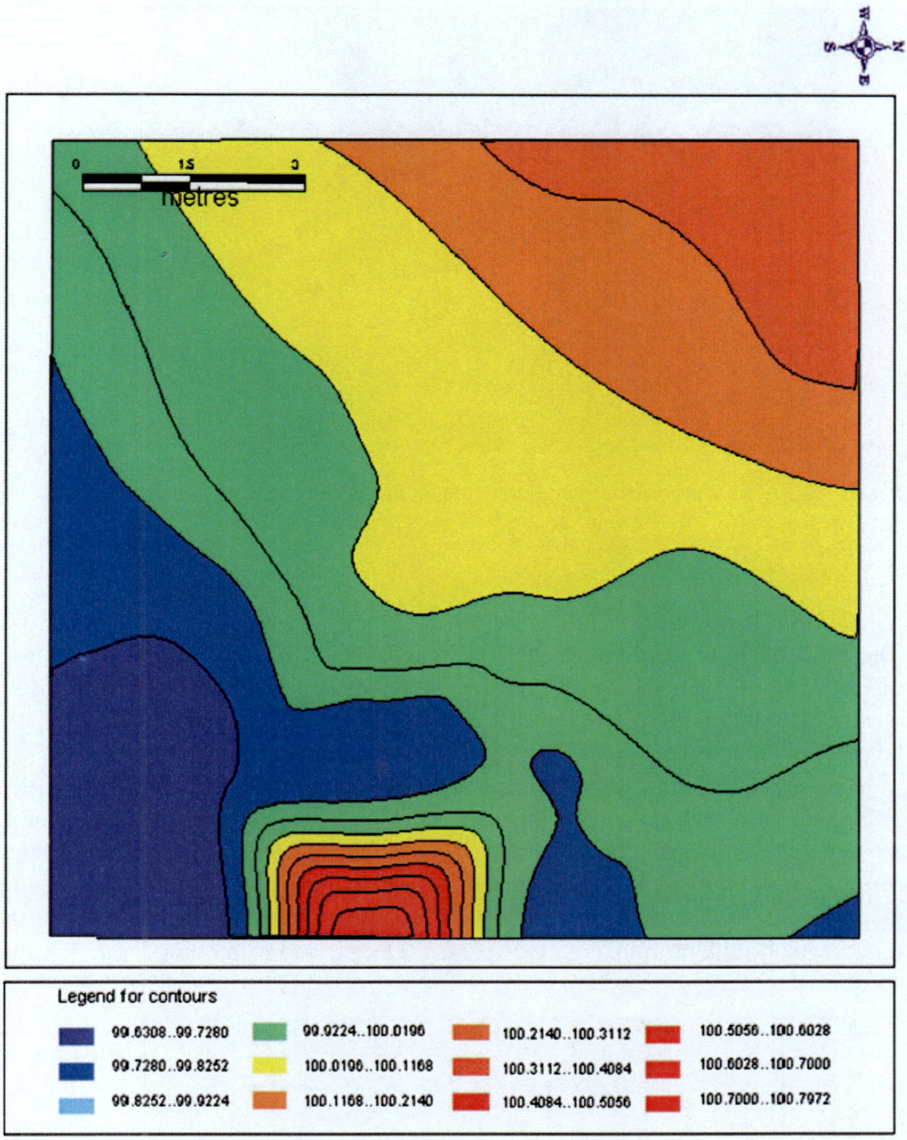

Fig. 9 Detail of the micro-level elevation contours of Site 1 (cf. Fig. 8)

reduce the risk of snakes, scorpions and wasps. Due to this constant clearing of vegetated areas in and around the constructions, and the concomitant erosion, brings so much soil out with water (Fig. 10) that portions of macadamised roads begin to look like mud-tracks after the rains cease. All-in-all campus development without regards to protecting ecosystem services has set the campus on course of incremental eco-degradation.

Fig. 10 Soil erosion accompanying rainwater runoff in the study area

3.2 Vegetation

The flora of those portions of the study area, which has not yet been seriously disturbed in the campus, constitutes diverse habits and habitats. Herbs and shrubs coexist with trees (Fig. 11); their diversity being adequate to enable the study area being classified as woodlands [4].

The bulk (62%) of the green cover is provided by the species *Azadirachta indica*, *Pongamia sp.*, and *Borassis*. About 34% species are associated with the classification *scrub jungle*, while 17% come under wastelands. That the natural vegetation has been seriously interfered with is menifest in the presence of species such as *Lucas aspera*, *Dodonea viscosa*, and *Tridax procumbens*. They now form as much as 14% of the total vegetation. Unfortunately, the species selected so far to compensate the loss

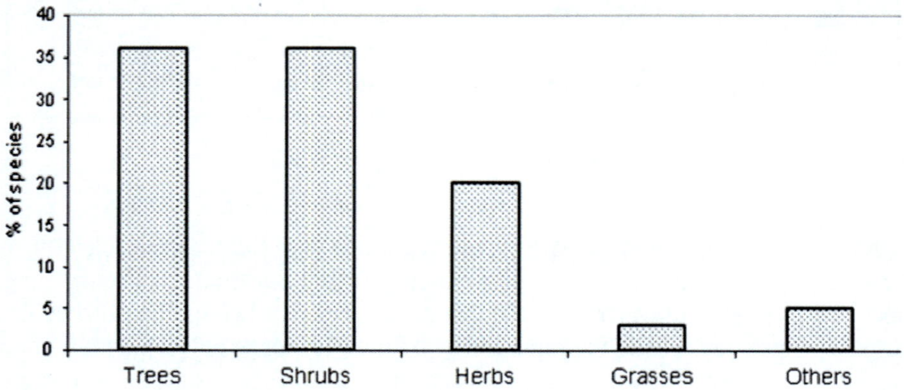

Fig. 11 Habitat distribution of the vegetation

of trees for construction activities, are predominantly *Acacia sp.* and *Eucalyptus sp.* While these species are ideal for restoring or utilizing seriously degraded lands [5, 6], they are not at all appropriate for the study area. Since the study area has had good soil it could easily have supported less dominant and more varied flora.

3.3 Other Features

In order that further ecodegradation of the campus is halted, there is an imperative necessity to regulate further development along ecologically compatible lines as also to restore the degraded areas to the maximum extent possible. There is also an urgent need to device an integrated and ecologically compatible strategy for handling the liquid and solid wastes that are being generated at the campus. These wastes have already attained increasingly menacing proportions, more so due to a rapid increase in the campus population in the recent years.

3.4 Imperatives

The management of the study area needs to take extensive measures, on a priority basis, to improve ground water recharge, contain soil erosion, and treat solid and liquid wastes. On the basis of the findings presented above, the following imperatives emerge:

(i) *Fecilitating groundwater recharge*: Numerous sites exist in the campus where groundwater recharge can be helped through better management of the runoff. Besides encouraging natural vegetation in all open areas, contour bunds, sand traps and other mechanical measures should be used to delay run-off and promote ground water recharge. Many of these measures require little expenses.

(ii) *Implementing rain water harvesting*: Huge spans of rooftop area lie unutilized in the campus at present. With easily implementable and fairly inexpensive measures rain water can be collected at the roof-tops and diverted into cisterns. Further runoff can be disposed in percolation pits to not only help in recharging the ground water aquifers but also in reducing the run-off losses. The harvested rainwater can greatly supplement the water requirement of the campus dwellers. But, strangely, no rooftop rainwater harvesting is carried out. It is a typical situation, encountered in most institutions, where concerns for environment are liberally expressed but are rarely seen translated into practice.

(iii) *Disposal of liquid waste*: The study area is ideal for wastewater treatment based on SHEFROL® technology [7–10] because of adequate land availability as well the long years of experience the university has in the successful implementation of this technology at pilot scale (Fig. 12). The SHEFROL® units can

Fig. 12 One of the real life sewage treatment plants based on the SHEFROL technology treating the wastewater generated by a segment of the Pondicherry University campus. All SHEFROL units have similarly green and aesthetically pleasing ambience

be integrated with downstream processes for phytomass utilization [11–18] to upgrade them to biorefineries [19–24].

(iv) *Disposal of solid waste*: sorting out solid waste at site and conversion of biodegradable waste into either compost/vermicompost.

(v) More careful choice of species to be planted for ecorestoration.

Acknowledgements Authors profusely thank Thiru T. Senthil Murugan for extensive help. SAA gratefully acknowledges Ministry of Water Resources (Jal Shakti Mantralaya), Government of India, for project grant 23/30.2004-R&D/957-968, and the Indian National Science Academy, New Delhi, for Emeritus Professorship.

References

1. Abbasi T, Abbasi SA (2010) Remote sensing, GIS and wetland management, Discovery Publishing House, New Delhi, pp vii+411
2. Chari KB, Abbasi SA (2005) Map viewer: a new add-on tool for geographic information system. Inf Technol J 4:469–475
3. Abbasi SA, Abbasi T (2022) R&D-cum-action and awareness project on rainwater harvesting, groundwater recharge and wastewater reuse. Completion report 23/30.2004-R&D/957-968, vol 1&2, pp vi +738
4. Oliver R (2006) Woodlands, Harper Collins, London. ISBN 9780007202447, pp xi+335
5. Abbasi SA, Vinithan S (1997) Ecological impacts of eucalypts—myths and realities. Indian For 123(8):710–739
6. Abbasi SA, Vinithan S (1999) Ecological impacts of *Eucalyptus tereticorns-globulus* (eucalyptus hybrid) plantation in a mining area. Indian For 125:163–184
7. Tabassum-Abbasi, Patnaik P, Tauseef SM, Abbasi SA (2022) SHEFROL® bioreactor enhances the ability of pistia (*Pistia stratiotes*) in the phytoremediation of greywater. Int J Environ Anal Chem 102:1502–1511. https://doi.org/10.1080/03067319.2020.1738420.

8. Tabassum-Abbasi, Patnaik P, Abbasi SA, Abbasi T (2022) Feasibility of a novel (SHEFROL) technology in pre-treating eatery wastewater at pilot scale. Water 14:3789. https://doi.org/10.3390/w14223789
9. Tabassum-Abbasi, Patnaik P, Abbasi SA (2022) Process design and assessment of the performance of three macrophytes in a biorefinery polishing partly treated sewage in novel SHEFROL bioreactors. Processes 10:2350. https://doi.org/10.3390/pr10112350
10. Bhat MA, Tabassum-Abbasi, Abbasi T, Abbasi SA (2022) An inexpensive phytoremediation system for treating 50,000 L/day of sewage. Int J Phytorem 1–13. https://doi.org/10.1080/15226514.2022.2125497
11. Bhat MA, Abbasi T, Abbasi SA (2021) A pilot plant of SHEFROL phytoremediation technology for treating greywater of a typical Indian village. Ind J Environ Prot 41:668–672. ISSN: 0253-7141
12. Hussain N, Abbasi SA (2018) Efficacy of the vermicomposts of different organic wastes as "clean" fertilizers: state-of-the-art. Sustainability 10:1–63. https://doi.org/10.3390/su10041205
13. Hussain N, Abbasi T, Abbasi SA (2016) Vermicomposting transforms allelopathic parthenium into a benign organic fertilizer. J Environ Manage 180c:180–189
14. Tauseef SM, Tabassum-Abbasi, Patnaik P, Abbasi T, Abbasi SA (2021) A novel high-rate vermicomposting machine for downstream processing of biorefinery waste and other forms of organic solid waste. Biomass Conv Biorefinery 1–10. https://doi.org/10.1007/s13399-021-01522-w
15. Banupriya D, Tabassum-Abbasi, Abbasi T, Abbasi SA (2022) Rapid, clean, and sustainable bioprocessing of toxic weeds into benign organic fertilizer. Agriculture 12:1511. https://doi.org/10.3390/agriculture12101511
16. Abbasi SA, Tabassum-Abbasi, Patnaik P (2021) Biomimetically generated nanoparticles in boosting the performance of microbial fuel cells. Nature Environ Poll Tech 20:1881–1886. https://doi.org/10.46488/NEPT.2021.v20i05.003
17. Tabassum-Abbasi, Abbasi T, Abbasi SA (2019) Microbial fuel cells as source of clean energy—potential and pitfalls. Nature Environ Poll Tech 18:789–797
18. Tabassum-Abbasi, Khamrang C, Abbasi T, Abbasi SA (2022) The fecundity and the vermicomposting efficiency of three generations of four earthworm species in high-rate vermireactors fed with the pernicious aquatic weed salvinia. Environ Sci Poll Res 1–12. https://doi.org/10.1007/s11356-022-24085-3
19. Tabassum-Abbasi, Patnaik P, Rahi R, Abbasi SA (2022) A circular biorefinery-integrating wastewater treatment with the generation of an energy precursor and an organic fertilizer. Sustainability (Switzerland) 14:5714. https://doi.org/10.3390/su14095714
20. Tabassum-Abbasi, Patnaik P, Abbasi SA (2022) A closed loop wastewater biorefinery based on the recently introduced shefrol reactor and two freely available macrophytes. Clean Soil Air Water 1–12. https://doi.org/10.1002/clen.202100287
21. Tabassum-Abbasi, Abbasi T, Abbasi SA (2018) Role of HAZOP in assessing risk of accidents in chemical process industries: capability and lacunae. Int J Eng Sci Math 7:482–487
22. Abbasi T, Abbasi SA (2010) Renewable energy sources: their Impact on global warming and pollution. PHI Larning, New Delhi, pp x + 306
23. Tauseef SM, Rashtchian D, Abbasi T, Abbasi SA (2011) A method for simulation of vapour cloud explosions based on computational fluid dynamics (CFD). J Loss Prev Proc Ind 24:638–647
24. Abbasi SA (1989) Atomic absorption spectrometric and spectrophotometric trace analysis of uranium in environmental samples with Np-methoxyphenyl-2-furylacrylohydroxamic acid and 4-(2-pyridylazo) resorcinol. Int J Environ Anal Chem 36:163–172
25. Chari KB, Abbasi SA (2005) A study on the fish fauna of oussudu–a rare freshwater lake of south India. Int J Environ Stud (Taylor & Francis) 62:137–145
26. Siddiqui NA, Tauseef SM, Abbasi SA, Rangwala AS (2018) Advances in fire and process safety, Springer transactions in civil and environmental engineering. Springer, Singapore, pp xi+336. https://doi.org/10.1007/978-981-10-7281-9

27. Tabassum-Abbasi, Patnaik P, Abbasi SA (2019) Ability of Indian pennywort Bacopa monnieri (L.) Pennell in the phytoremediation of sewage (greywater). Environ Sci Pollut Res 27(6):6078–6087. https://doi.org/10.1007/s11356-019-07259-4
28. Abbasi SA, Tabassum-Abbasi, Ponni G, Tauseef SM (2019) Potential of joyweed *Alternanthera sessilis* for rapid treatment of domestic sewage in SHEFROL® bioreactor. Int J Phytoremed 21(2):160–169.https://doi.org/10.1080/15226514.2018.1488814
29. Tabassum-Abbasi, Patnaik P, Abbasi SA (2021) Screening of twenty-three common terrestrial plants for their possible use as phytoremediators of greywater in SHEFROL®. Taiwan Water Conserv 69(2):24–35https://doi.org/10.6937/TWC.202106/PP_69(2).0004
30. Tabassum-Abbasi, Patnaik P, Abbasi SA (2022) Use of the ornamental plant *sadabahaar* (*Catharanthus roseus*) in beautifying and enhancing the SHEFROL® greywater treatment unit, with concomitant increase in the level of treatment, Taiwan Water Conserv. In Press. https://doi.org/10.6937/TWC.20xxxx/PP_xx(x).0001

The Weed *Plantago* Major is an Effective Bioagent in Achieving Rapid Treatment of Sewage in SHEFROL® Bioreactor

Ashraf Zainabi, Arshad Hussain, Gulzar A. Ganie, Obeida Ashraf, and Tabassum-Abbasi

Abstract The potential of the weedy terrestrial herb *Plantago major* which is variously called "great plantain", "white man's footprint", "way bread", and "broad leaf plantain", has been explored in serving as a bioagent in the treatment of sewage in the recently patented and trademarked Sheet-flow Root Level (SHEFROL®) bioreactors. It was seen that despite being used without the support of soil or gravel, *P. major* was as effective in sewage treatment as several aquatic weeds that have been explored earlier. It was able to remove suspended solids (SS), 5-day biological oxygen demand (BOD_5), chemical oxygen demand (COD), total Kjeldahl nitrogen (TKN), phosphorus (P), and total dissolved solids (TDS) to the extents of 97.1, 79.5, 74.9, 40.5, 26.7, and 3.1%, respectively. The weed was not only able to survive in SHEFROL® but seemed to thrive, remaining green and vigorous throughout the experiment lasting two months.

A. Zainabi
Department of Zoology, Model Government Degree College Charar-i-Sharief, Budgam, Jammu and Kashmir 191 112, India

A. Hussain
Department of Botany, Model Government Degree College Charar-i-Sharief, Budgam, Jammu and Kashmir 191 112, India

G. A. Ganie
Department of Zoology, Government Degree College Pulwama (Boys), Pulwama, Jammu and Kashmir 192 301, India

O. Ashraf
House Number 10, Mir Mohalla Gowharpora, Chadoora, Jammu and Kashmir 191113, India

Tabassum-Abbasi (✉)
School of Engineering, University of Petroleum and Energy Studies, Bidholi, Dehradun 248 007, India
e-mail: Tab.abbasi@gmail.com

N. A. Siddiqui et al. (eds.), *Advances in Waste Management*, Lecture Notes in Civil Engineering 301, https://doi.org/10.1007/978-981-19-7506-6_15

1 Introduction

A novel bioreactor, acronymed SHEFROL®, has been recently introduced by Abbasi and co-workers [1, 2]. It enables as effective treatment of sewage as other phytoremediation systems such as constructed wetlands are able to achieve, but does so much more efficiently and inexpensively [3, 4]. The technology revolves around the use of short-sized plants which may be free-floating aquatics like salvinia (*Salvinia molesta*) and pistia (*Pistia stratiotes*), or rooted marshland species like four-leaf clover (*Marsilea quadrifolia*) or even pure terrestrials. The plants are stocked in narrow channels, of which the dimensions have been optimized to enable the most effective plant-sewage contact while minimizing short-circuiting and maximizing agitation of the passing sewage. The narrowness of the channels also enables the plants to be stocked without the need for any soil/gravel or scaffold [2, 4]. The system is so designed that sewage passes through these channels as a "sheet" deep enough to only reach up to the plant roots. It is this "sheet-flow root level" (SHEFROL) hydrology that controls much of the functioning of the SHEFROL® reactors which has prompted the name of the technology. It has since been trademarked [5].

After the efficiency of the technology had been established at the bench scale, it has been extensively validated at the pilot plant scale and in real-life systems [3, 6]. A large number of species have also been explored for their compatibility with SHEFROL® [2, 7–16].

This paper describes a study in which the efficacy of *Plantago major* (Figs. 1 and 2) has been explored for possible use as a bioagent for treating sewage in SHEFROL® reactors. *P. major* is a terrestrial herb of European/Central Asian origin which has widely naturalized in most of the world's regions. It can grow about 15 cm tall but the average sizes of adult *P. major* can vary from habitat to habitat and region to region. Its ability to create up to 20,000 seeds per plant, and to wind-pollinate, has helped its spread, turning it into a weed. It finds limited application as a medicinal, ornamental, and food plant [17–19] but the quantities of *P. major* that are so utilized are insignificant in comparison to the quantities that are generated. If the efficacy of *P. major* as a phytoremediator in SHEFROL® can be established, the weed can be put to much greater use than it is at present.

2 Materials and Method

SHEFROL® reactor was fabricated and operated essentially as detailed earlier [2, 13, 14]. The SHEFROL® channels were 1.5 m long, 0.15 m wide, and 0.1 m deep. Sewage was served on the channels from a 60 L settling-cum-flow-equalization tank. A representative reactor showing an empty channel and the one planted with *P. major* is shown in Fig. 3.

The manner of drawing samples, analyzing them for various variables, and assuring quality control was also essentially as detailed earlier [2].

Fig. 1 A typical spread of *P. major*

Fig. 2 Close up of a *P. major* plant

Fig. 3 The SHEFROL® set up with *P. major* as the main bioagent

3 Results and Discussion

As has been seen earlier in case of all SHEFROL® systems run on different species of bioagents, this system also started treating sewage to a significant extent, achieving about 40% reduction in COD, and proportionate reduction in other contaminants from the first 24 h of start. The extent of treatment improved with each passing day till steady state was achieved in 10 days. Thereafter, the extent of treatment hovered within a variation of 0–10% from day to day. In other words, the reactor had achieved a steady state. All figures of treatment achieved in this report belong to the steady-state SHEFROL® operation.

The range of concentrations of various contaminants in the influent sewage, and the range present in the effluent are summarized in Table 1. The extent of treatment achieved in terms of percent reduction is also shown in Table 1.

As may be seen, suspended solids (SS) were almost totally removed. The reduction in the 5-day biological oxygen demand (BOD_5) was in the range of 73.1–79.5%. The chemical oxygen demand (COD) was also removed very substantially, reaching up to 76.9%. The extent of average removal of total Kjeldahl nitrogen (TKN) and phosphorous (P) – 37.6% and 24.1% – matched that reported generally from constructed wetlands [20], albeit at much higher HRTs than the mere 7.5 h HRT of this SHEFROL®. In the like manner, the removal of other contaminants in this SHEFROL® was on a par with that achieved by constructed wetlands [21], but

Table 1 Influent concentrations and extent of reduction achieved by SHEFROL® in them at a hydraulic retention time (HRT) of 7.5 h after the reactor attained steady state

Variable	Range of		
	Influent concentration, mg/L	Effluent concentration, mg/L	Reduction achieved, %
SS	6.8–7.4	0.2–0.7	90.5–97.1
BOD_5	108–137	25–36	73.1–79.5
COD	223–252	56–74	65.4–74.9
TKN	6.9–7.4	4.4–4.7	34.8–40.5
P	1.2–1.5	0.9–1.1	21.4–26.7
TDS	33.2–42.4	31.6–39.2	4.9–7.5

at a much shorter HRT, indicating much higher efficiency of SHEFROL® despite the simplicity and the inexpensiveness of the reactor.

As may be seen from Fig. 1, *P. major* remained vibrant and healthy in SHEFROL® channels, showing no signs of stress. It is, thus, seen that *P. major* can be added to the repertoire of other screened species of macrophytes that have proved effective as bioagents in SHEFROL® systems. After use in SHEFROL, the plant can be converted to energy precursors and/or organic fertilizers upgrading SHEFROL® units into biorefineries [22–33].

4 Mechanism of Sewage Treatment in SHEFROL®

Sewage treatment occurs in SHEFROL®, and is expedited, by a combination of factors involving the reactor hydrology and a large variety of physical, chemical, and biological influences. The roots of the plants play the most important of roles while the geometry and the operation protocol of the reactor facilitate those roles. These aspects have been reviewed elsewhere [2, 4, 34].

5 Summary and Conclusion

The efficiency of the perennial weedy herb *Plantago major* as a bioagent for treating sewage in sheet-flow-root-level (SHEFROL®) bioreactors was explored. The terrestrial species was seen to thrive in soil-less and gravel-free SHEFROL® channels, treating sewage while utilizing the nutrients present in the sewage. The macrophyte was able to remove suspended solids (SS), 5-day biological oxygen demand (BOD_5), chemical oxygen demand (COD), total Kjeldahl nitrogen (TKN), phosphorus (P), and total dissolved solids (TDS) to the extents of 97.1, 79.5, 74.9, 40.5, 26.7, and 3.1%, respectively. The weed was not only able to survive in SHEFROL®, but also seemed to thrive, remaining green and vigorous throughout the experiment lasting two months.

This level of performance of SHEFROL® was comparable to the one achieved with constructed wetlands but SHEFROL® achieves it at a much faster rate and lesser costs per litre of sewage treatment.

Acknowledgements Thanks to Prof. Gulzar Ahmad Ganie Government Degree College Pulwama, for bearing the cost of the setup. Thanks to the Student of the Department of Zoology, Model Government Degree College, Chararisharief, for collecting wastewater from a nearby village sewer channel.

References

1. Bhat MA, Abbasi T, Abbasi SA (2018) Soil-less use of aquatic macrophytes in wastewater treatment and the novel SHEFROL® bioreactor. In: Siddiqui NA, Tauseef SM, Kansal B (eds) Advances in health and environment safety, Springer transactions in civil and environmental engineering, Chapter 30, pp 297–316. https://doi.org/10.1007/978-981-10-7122-5_30
2. Abbasi SA, Tabassum-Abbasi, Ponni G, Tauseef SM (2019) Potential of joyweed *Alternanthera sessilis* for rapid treatment of domestic sewage in SHEFROL® bioreactor. Int J Phytorem 1–10. https://doi.org/10.1080/15226514.2018.1488814
3. Bhat MA, Abbasi T, Abbasi SA (2021) A pilot plant of SHEFROL® phytoremediation Technology for treating greywater of a typical Indian village. Ind J Environ Pollut 41:668–672. ISSN: 0253-7141
4. Bhat MA, Tabassum-Abbasi, Abbasi T, Abbasi SA (2022) An inexpensive phytoremediation system for treating 50,000 L/day of sewage. Int J Phytorem 1–13. https://doi.org/10.1080/15226514.2022.2125497
5. Abbasi SA (2021) Certificate of registration of trade mark, number 4735305, Registrar of Trademarks, Intellectual Property India, Government of India
6. Bhat MA, Abbasi T, Abbasi SA (2020) A household-scale on-site embodiment of the novel bioreactor SHEFROL® for treating greywater. Taiwan Water Conserv 68:26–35. https://doi.org/10.6937/TWC.202006/PP_68(2).0003
7. Abbasi SA, Tauseef SM (2018) A system for rapid and inexpensive treatment of sewage using the weed false daisy (*Eclipta prostrata*). Water Environ J. https://doi.org/10.1111/wej.12355
8. Abbasi SA, Tauseef SM (2018) Use of the terrestrial weed *Alternanthera ficoidea* in treating greywater in soil-less SHEFROL® bioreactors. Water Sci Technol (IWA) 77:2005–2013. https://doi.org/10.2166/wst.2018.093
9. Abbasi SA, Tauseef SM (2019) Rapid treatment of greywater (household sewage) by terrestrial weed *Achyranthes aspera* in SHEFROL® reactors. Environ Prog Sustain Energy 38:467–476. https://doi.org/10.1002/ep.12994
10. Abbasi SA, Ponni G, Tauseef SM (2018) *Marsilea quadrifolia*: a new bioagent for treating wastewater. In: Water, air soil pollution, vol 229. Springer, pp 1–8
11. Abbasi SA, Ponni G, Tauseef SM (2018) Proficiency of brahmi (Indian pennywort) *Hydrocotyle asiatica* in the one-pot secondary and tertiary treatment of sewage. Nat Environ Pollut Technol 17:603–609
12. Abbasi SA, Hussain N, Tauseef SM, Abbasi T (2018) A novel flippable units vermireactor train system–FLUVTS–for rapidly vermicomposting paper waste to an organic fertilizer. J Clean Prod 198:917–930. https://doi.org/10.1016/j.jclepro.2018.07.040
13. Tabassum-Abbasi, Patnaik P, Tauseef SM, Abbasi SA (2022) SHEFROL® bioreactor enhances the ability of pistia (*Pistia stratiotes*) in the phytoremediation of greywater. Int J Env Anal Chem, 102:1502–1511. https://doi.org/10.1080/03067319.2020.1738420.

14. Tabassum-Abbasi, Patnaik P, Abbasi SA (2022) A closed loop wastewater biorefinery based on the recently introduced SHEFROL reactor and two freely available macrophytes. Clean–Soil, Air, Water, 1–12. https://doi.org/10.1002/clen.202100287
15. Tabassum-Abbasi, Patnaik P, Abbasi SA (2021) Screening of twenty-three common terrestrial plants for their possible use as phytoremediators of greywater in SHEFROL®. Taiwan Water Conserv 69:24–35. https://doi.org/10.6937/TWC.202106/PP_69(2).0004
16. Tabassum-Abbasi, Patnaik P, Abbasi SA (2021) Use of the ornamental plant sadabahaar (*Catharanthus roseus*) in beautifying and enhancing the SHEFROL® greywater treatment unit, with concomitant increase in the level of treatment. Taiwan Water Conserv 69:73–87
17. Keivani M, Mehregan I, Albach DC (2021) Evaluating morphological diversity among *Plantago major L.* populations and influence of ecological variables. Biologia 76:1127–1139
18. Adom MB, Taher M, Mutalabisin MF, Amri MS, Abdul Kudos MB, Wan Sulaiman MWA, Sengupta P, Susanti D (2017) Chemical constituents and medical benefits of *Plantago major*. Biomed Pharmacother 96:348–360. https://doi.org/10.1016/j.biopha.2017.09.152
19. Samuelsen AB (2000) The traditional uses, chemical constituents and biological activities of *Plantago major L.* A Rev J Ethnopharmacol 71:1–21
20. Brix H, Schierup HH, Arias CA (2007) Twenty years experience with constructed wetland systems in Denmark—What did we learn? Water Sci Technol 56:63–68
21. Jozwiakowski K, Bugajski P, Mucha Z, Wojcik W, Jucherski A, Nastawny M, Siwiec T, Mazur A, Obroslak R, Gajewska M (2017) Reliability and efficiency of pollution removal during long-term operation of a one-stage constructed wetland system with horizontal flow. Separ Purif Technol 187:60–66
22. Abbasi SA, Nayeem-Shah M, Tasneem Abbasi (2015) Vermicomposting of phytomass: limitations of the past approaches and the emerging directions. J Clean Prod 93:103–114
23. Abbasi T, Abbasi SA (2010) Factors which facilitate waste water treatment in presence of aquatic weeds-the mechanism of the weed's purifying action. Int J Environ Stud (Rutledge) 67:349–371
24. Hussain N, Abbasi SA (2018) Efficacy of the vermicomposts of different organic wastes as "clean" fertilizers: state-of-the-art. Sustainability 10:1–63. https://doi.org/10.3390/su10041205
25. Hussain N, Abbasi T, Abbasi SA (2016) Vermicomposting transforms allelopathic parthenium into a benign organic fertilizer. J Environ Manage 180c:180–189
26. Tauseef SM, Tabassum-Abbasi, Patnaik P, Abbasi T, Abbasi SA (2021) A novel high-rate vermicomposting machine for downstream processing of biorefinery waste and other forms of organic solid waste. Biom Conv Biorefinery 1–10. https://doi.org/10.1007/s13399-021-01522-w
27. Tabassum-Abbasi, Patnaik P, Rahi R, Abbasi SA (2022) A circular biorefinery-integrating wastewater treatment with the generation of an energy precursor and an organic fertilizer. Sustainability (Switzerland) 14:5714. https://doi.org/10.3390/su14095714
28. Abbasi T, Mishra PK, Shreevidhya R, Abbasi SA (2022) Exploration of organic wastes and problematic weeds as absorbents for the removal of oil and grease from greywater. Ind J Environ Prot 42:631–636
29. Tabassum-Abbasi, Patnaik P, Abbasi SA, Abbasi T (2022) Feasibility of a novel (SHEFROL) technology in pre-treating eatery wastewater at pilot scale. Water 14:3789. https://doi.org/10.3390/w14223789
30. Tabassum-Abbasi, Patnaik P, Abbasi SA (2022) Process design and assessment of the performance of three macrophytes in a biorefinery polishing partly treated sewage in novel SHEFROL bioreactors. Processes 10:2350. https://doi.org/10.3390/pr10112350
31. Tabassum-Abbasi, Abbasi T, Abbasi SA (2019) Microbial fuel cells as source of clean energy—potential and pitfalls. Nat Environ Pol Tech 18:789–797
32. Abbasi SA, Tabassum-Abbasi, Patnaik P (2021) Biomimetically generated nanoparticles in boosting the performance of microbial fuel cells. Nat Environ Pol Tech 20:1881–1886. https://doi.org/10.46488/NEPT.2021.v20i05.003

33. Tauseef SM, Rashtchian D, Abbasi SA (2011) CFD-based simulation of dense gas dispersion in presence of obstacles. J Loss Prev Process Ind 24:371–376
34. Tabassum-Abbasi, Abbasi T, Abbasi SA (2018) Role of HAZOP in assessing risk of accidents in chemical process industries: capability and lacunae. Int J Eng Sci Math 7:482–487

Study on Coconut Shell Aggregate Behavior and the Use of Super Absorbent Polymer as an Internal Curing Agent

M. Tamilselvi, A. K. Dasarathy, S. Ponkumar Ilango, and V. Karthika

Abstract *One* alternative for environmentally friendly construction is to utilize discarded materials. An experimental examination was carried out in this paper to measure the compressive strength, flexural strength, load, and displacement of concrete specimens utilizing coconut shell waste as coarse aggregate after it was crushed to make lightweight concrete. The current research has used coconut shell waste in concrete as a partial replacement for coarse aggregate in various percentages, i.e., at 25% coconut shell with 0%SAP, 25% Coconut shell with 0.5% of SAP, 25% coconut shell with 1% of SAP, 25% coconut shell with 1.5% of Super absorbent polymer, 25% coconut shell with 2% of a super absorbent polymer. Various material tests were performed, including sieve analysis, specific gravity, water absorption, and bulk density test. The concrete specimens' further tests of compressive strength, flexural strength, Deflection, Stress, Strain, and Modulus of elasticity, as well as crack breadth, will be determined after 28 days of curing. The amount of super absorbent polymer that should be used on the compressive strength.

Keywords Coconut shell · Super absorbent polymer · Light weight concrete · Internal curing · Durability · Sustainable development

M. Tamilselvi (✉) · V. Karthika
Civil Engineering Department, S. A. Engineering College (Autonomous), Chennai, India
e-mail: tamilselvi_05@yahoo.com

A. K. Dasarathy
Jain Deemed to be University, Bangalore, India

S. P. Ilango
Architectural Department, Dr. M. G. R. Educational and Research Institute University, Chennai, India

N. A. Siddiqui et al. (eds.), *Advances in Waste Management*, Lecture Notes in Civil Engineering 301, https://doi.org/10.1007/978-981-19-7506-6_16

1 Introduction

Concrete is the extensively used primary structural fabric with inside the international Today. The call for to make this fabric lighter has been the problem of observe that has challenged scientists and engineers alike. The undertaking in creating a light-weight concrete is lowering the density whilst preserving electricity and without adversely affecting cost. Introducing new aggregates into the combination pure layout is a not unusual place manner to decrease a concrete's density. Normal concrete is made up of four ingredients: cement, crushed stone, river sand, and water.

Crushed stone and sand are typically replaced with lightweight aggregates. Lightweight concrete is typically created by incorporating natural or synthetic lightweight aggregates into a concrete mixture or by entraining air into a concrete mixture. Coconut is thought to have originated in South East Asia. India is the third-largest producer of coconuts. However, it's also the primary contributor to the nation's pollutants hassle as a stable waste withinside the shape of shells, which includes an annual manufacturing of about 3.20 million tons.

Coconut shell represents greater than 65% of the home waste volume. Coconut shell, which provides critical disposal troubles for the neighborhood environment, is an abundantly to be had as an agricultural waste from neighborhood coconut industries. In developing countries where there is an abundance of agricultural and industrial waste, these wastes can be used as potential or replacement material in the construction industry. This will have the dual benefit of lowering the cost of construction materials while also serving as a means of waste disposal.

Anwar et al. [1]: They did research. M Waste coconut shell become used to update herbal coarse aggregates at zero percentage, five percentage, 10%, 20%, 30%, 40%, and 50% via way of means of weight in a 20-grade concrete. Sixty-three (63) cubes have been forged in total, and their compressive power become measured after seven, fourteen, and 28 days. As the proportion of alternative rose, the compressive power of concrete decreased. The concrete compositions have been placed to the test. At 28 days, the compressive power of the concrete combinations become evaluated and as compared to that of well-known concrete. The findings found out that Coconut Shell Concrete (CSC) may be hired withinside the production of lightweight concrete structures [1].

Deshmukh et al. [2]: They promoted the use of coconut shell as an aggregate in both low-cost dwellings and mass concrete structures made with a high-strength mix. The report also looks at how concrete might be used in building in a sustainable way. The use of coconut shell into concrete aids in the production of lightweight concrete. In the current investigation, nominal mix M20 will be replaced with various combinations of natural material and coconut shell in proportions of 0%, 20%, 40%, and 50%. For each percent replacement, twelve sample specimens will be made. Each cube specimen's compressive strength behavior will be studied for days [2].

Krishnamurthy et al. [3]: Internal curing was studied as a supplement to traditional curing in conventional concrete. Internal curing was accomplished by introducing super absorbent polymer (SAP) in proportion to the weight of the sand. SAP's internal water curing substantially reduced early-age chemical shrinkage while also increasing concrete's compressive strength [3].

Ankit Dubey et al. [4]: The test on various strengths of hardened SAP-induced concrete was conducted and the results were compared to standard M25 grade concrete. The strength of concrete is significantly influenced by this form of curing. This is also possible because to SAP's crack-sealing mechanism, which aids in the prevention of cracks in concrete caused by the hydration process. He looked into how using super absorbent polymers (SAP) in concrete can improve a variety of concrete qualities [4].

Sequeira and Pai [5]: The problem that hinders traditional water curing methods from being successfully used in HSC was investigated. As a result, we turn to alternative materials such as lightweight aggregate or Super absorbent polymers, which can collect, hold, and deliver water as needed while simultaneously serving as aggregate [5].

Similar works carried by Duna and Musa [6], Shraddha and Vaniya [7] and Yerramala and Ramachandrudu [8]: The findings revealed that, in general, strength decreases as coconut shell content increases and increases as the curing duration lengthens [6–8].

Kang et al. [9] and Afolayam et al. [10]: They looked examined the effects of different ions in a cement-based solution on the water retention capacity and ion absorbency of a superabsorbent polymer (SAP) [9, 10].

2 Materials Used

Cement: In the field, conventional Portland cement of 53 grates, manufactured by the Coromandel Cement Company in accordance with IS 8112:1989, was utilized.

Coarse Aggregate (Crushed Stone)

It is decided to use crushed granite stone aggregate with a maximum size of 20 mm. Surface texture features of the aggregate as classified in IS: 383–1970, as coarse aggregate according to IS: 2386 (Part I) 1963.

Aggregate of Coconut Shells

This investigation employed a freshly discarded coconut shell gathered from a local oil mill. Because several species of coconut shell are processed simultaneously, the thickness of the shells varies from 2–8 mm. The shells are flaky and unevenly shaped after crushing.

Fine Aggregate (Sand)

The sand is taken from the bottom of a polar river. Before being used, the sand has been sieved through a 4.75 mm sieve.

Water

Whenever drinkable water is available on university grounds that is free of all impurities, it is used for the entire work.

3 Super Absorbent Polymer

Self-curing materials such as super absorbent polymer, for example, absorb water several times its volume and harden into a stiff gel. It has the ability to hold moisture for an extended period of time. Self-curing is a process that adds moisture to concrete to help the cement hydrate more effectively.

COCONUT SHELL

Design Mix

Various mixes were made to study the strength of coconut shell concrete, including 0% SAP with coconut shell (i.e., replacing coconut shell with coarse aggregate 50% by volume),0.5% SAP with 50%, 25% coconut shell, 1% SAP with 50%, 25% coconut shell, 1.5% sap with 50%, 25% coconut shell, 2% SAP with 50%, 25% coconut shell, and 2% SAP with 50%, 25% coconut shell. The coconut shell should be soaked for at least 24 h before usage. To develop a mix design (normal concrete of grade M-20) in the lab using the IS technique according to IS 10262–2009.

By weight

Water	Cement	FA	CA
191.16 kg	383.16 kg	574 kg	1149 kg

By Ratio

Water	Cement	FA	CA
0.5	1	1.5	3

4 Experimental Work

The concrete is mixed by hand. The time spent mixing is limited to roughly 15–20 min. After mixing, each blend was put through a slump cone test to see if it was workable.

Compressive Strength

The concrete was placed in a well-oiled mould with dimensions of 150X150X150mm. For each mix, six molds were utilized. A total of 60 cubes were cast. After 24 h, the specimen was de-moulded. The nominal cubes were immersed in the curing tank till the testing date arrived. For moist curing, SAP-infused cubes were left out in the open at room temperature. The specimen was evaluated using a compression testing equipment after it had been cured for a length of time. Compressive strength is calculated using a formula.

$$= P/A.$$

P = Load.

A = Cross sectional area.

Average Compressive Strength at 7 and 28 Days for 25% CS.

M20	Average compressive strength at 7 days	Average compressive strength at 28 days
Nominal	18.162	28.12
0% SAP + 25% CS	20.50	30.76
0.5% SAP + 25% CS	21.14	32.59
1% SAP + 25% CS	18.93	27.63
1.5% SAP + 25% CS	17.07	25.47
2% SAP + 25% CS	15.46	23.70

Split Tensile Strength

For each mix proportion, a cylinder specimen with a height of 300 mm and a diameter of 150 mm must be cast. A total of 60 cylinders were cast. After 24 h, the specimen was de-moulded. The nominal cylinders were immersed in the curing tank till the testing date arrived. For moist curing, SAP stated that cylinders were left out in the open at room temperature. The specimen was evaluated using a compression testing equipment after it had been cured for a length of time. To calculate split tensile strength, use the formula below.

$$= 2p/\pi \text{ D L of cylinder}$$

M20	Average split tensile strength at 7 days	Average split tensile strength at 28 days
Nominal	1.05	2.17
0% SAP + 25% CS	1.40	2.11
0.5% SAP + 25% CS	1.14	2.38
1% SAP + 25% CS	0.92	1.95
1.5% SAP + 25% CS	0.84	1.62
2% SAP + 25% CS	0.66	1.49

Average split tensile Strength at 7 and 28 days.

TEST SET UP - COMPRESSIVE STRENGTH

Flexural Beam Strength Test:

For each mix proportion, a prism specimen with a length of 700 mm and a width of 150 mm should be cast. After 24 h, the specimen was de-moulded. The nominal prisms were immersed in the curing tank till the testing day arrived. For moist curing, SAP-infused prisms were left out in the open at room temperature. The specimen was evaluated utilizing a Universal Testing Machine after curing for a length of time (UTM). To calculate split tensile strength, use the formula below

$$= PL/BD^{\overline{2}}$$

Average flexural Strength at 7 and 28 Days for 25% CS.

M20	Average flexural strength at 7 days	Average flexural strength at 28 days
Nominal	2.7	3.66
0% SAP + 25% CS	2.21	3.31
0.5% SAP + 25% CS	2.87	3.85
1% SAP + 25% CS	2.08	3.34
1.5% SAP + 25% CS	1.76	3.16
2% SAP + 25% CS	1.64	2.98

5 Result and Discussion

The compressive strength test, split tensile strength test, and flexural strength test findings from the 7-day and 28-day tests are discussed here.

Compressive Strength

The compressive strength of nominal concrete with and without curing at intervals of 7 and 28 days is shown in the table and figure.

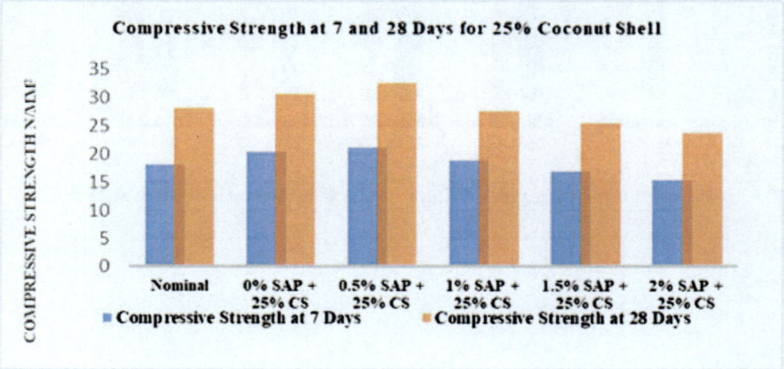

Split Tensile Strength Test

The split tensile strength for nominal concrete with curing and without curing at intervals of 7 days and 28 days is shown in the table and figure.

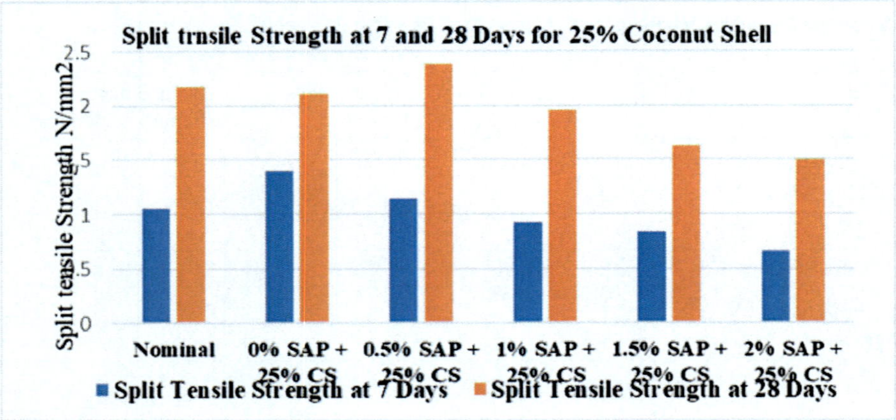

Flexural Strength Test

The flexural strength for nominal concrete with curing and without curing at intervals of 7 days and 28 days is shown in the table and figure.

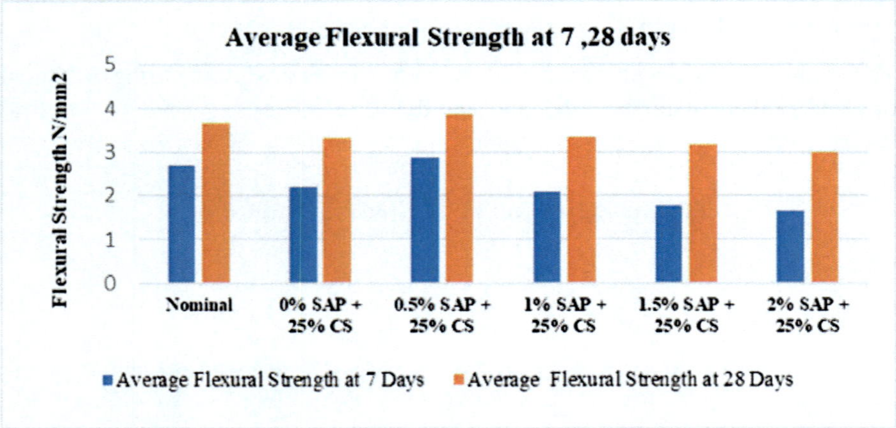

Maximum Strength Proportion

SL.NO	Type of beam	Load (KN)	Deflection (mm)
1	Conventional Beam	**65**	**5.2**
2	CS 25% + 0.5% SAP	**71**	**5.8**

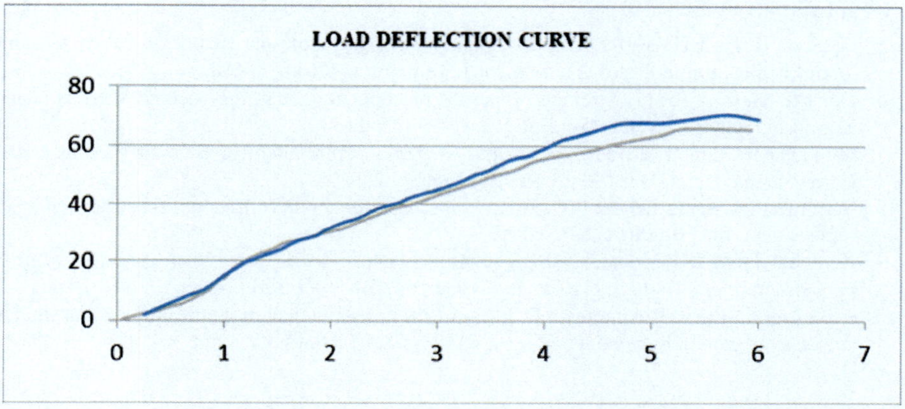

Conclusion

1. The use of super absorbent polymer in a study on coconut shell aggregate was shown to be effective in terms of both strength and cost efficiency.
2. The efficiency of ordinary concrete and ultra-absorbent polymer concrete made from coconut shell was compared.
3. The M20 concrete grade was utilized in both traditional and polymer concrete.
4. The conclusions are based on the examination of experimental results for compressive strength, split tensile strength, and flexural strength readings.
5. Type H concrete has a higher strength than ordinary concrete because it contains 25% coconut shell and 0.5% super absorbent polymer.
6. When compared to other self-curing agents such as super absorbent polymer, the specimen with 0.5% of self-curing agent had the maximum strength.
7. The compressive strength of the mix containing 0.5% SAP was almost identical to that of normal concrete.
8. Since a result of the test, it was determined that a sample containing 0.5% SAP is the perfect blend, as it has no effect on workability or compressive strength.
9. When the compressive strength of conventional concrete and self-curing concrete are compared, the self-curing concrete outperforms the conventional concrete.

References

1. Anwar A, Ahmad S, Ahmed SA (2016) Performance of waste coconut shell as partial replacement of natural coarse aggregate in concrete. Int J Sci Eng Res 7(8):1802–1809
2. Deshmukh N, Ganorkar P, Ronghe P, Pallewar P (2016) Behavior of Concrete by replacing coarse aggregate with Coconut shell. Int J Res Sci Eng 2(1):52–58
3. Krishnamoorthy D, Letcham, Prabhu G (2011) Use of super absorbing polymers as internal curing agent for conventional concrete with high W/C ratio. J Construct Build Mater 27:19–23
4. Dubey A (2019) Use of super absorbent polymers (SAPs) in concrete. Int J Res Eng Appl Sci 6(3):146–150
5. Sequeira K, Pai BHV (2015) Effect of small percentage additions of superabsorbent polymer on mechanical properties of concrete. Int J Sci Eng Res 6(6):683–688
6. Duna S, Musa A (2017) Strength properties of concrete containing coconut shell as coarse aggregate replacement. Int J Sci Eng Res 8(4):1125–1134
7. Shraddha R. Vaniya (2015) Experiments on partial replacement of coconut shell as coarse aggregate in concrete. Int J Innov Res Sci Technol 2(1):78–87
8. Yerramalaa A, Ramachandrudu C (2012) Properties of concrete with coconut shells as aggregate replacement. Int J Eng Invent 1(6):21–31
9. Kang SH, Hong SG, Moon J (2018) The effect of superabsorbent polymer on various scale of pore structure in ultra-high performance concrete. Construct Build Mater 172(10):29–40
10. Afolayan JO, Ejeh SP, Amartey YD, Idris A (2017) Mechanical properties of coconut shell as coarse aggregate in lightweight concrete. Asian J Sci Technol 08(09):5626–5633

Ecological and Economic Impacts on Biofuel Production

S. Sivarathnakumar, T. Adhitiyan, S. Gubendhiran, and R. Praveenkumar

Abstract This chapter addresses the issues and impact of biofuel production in environment and economic aspects of bioenergy. In case of biofuel production, biogas, biodiesel, bioethanol are the most predominant products produced which utilises food crops/grains, lignocellulosic biomass as raw materials. This leads to significant impact on environmental aspects like agricultural land usage, food to fuel competition, food security, greenhouse gas emission, trade of biofuel water and biodiversity. It also has a huge impact on the economic aspects like energy security, employment opportunities and also on human health. The feasibility of biofuel production can be made sustainable only if the environmental and economic aspects are addressed properly.

Keywords Biofuels · Ecological impacts · Economic impacts · Sustainability · LCA

1 Introduction

The twenty first century denotes the time of praiseworthy accomplishments accomplished in the area of science, innovation, data, innovative work in R&D alongside the upgrade in the way of life. Notwithstanding, on the opposite side of the coin is more regrettable by rising populace with expansion in energy interest. In 2011, practically 85% of the world energy request was fulfilled by the petroleum products. But the consuming of petroleum derivatives at a current rate is probably going to make a huge environmental crisis globally, such as, Green House Gas emission, acid rain, ozone depletion, global warming, climatic changes, water scarcity, etc. These crisis forces the researchers to search for new energy resources which has to be sustainable, non-polluting and renewable source of energy fuels [1]. The researcher

S. Sivarathnakumar (✉) · T. Adhitiyan · S. Gubendhiran
Department of Chemical Engineering, Arunai Engineering College, Tiruvannamalai, India
e-mail: sivrathna@gmail.com

R. Praveenkumar
Department of Biotechnology, Arunai Engineering College, Tiruvannamalai, India

N. A. Siddiqui et al. (eds.), *Advances in Waste Management*, Lecture Notes in Civil Engineering 301, https://doi.org/10.1007/978-981-19-7506-6_17

all throughout the world have investigated numerous alternative bioenergy resources such as biomass, biogas, bio-alcohol, bioethanol, vegetable oil and biodiesel.

As predicted by Henry Ford in 1025, "The fuel of the future is going to come from fruit like that shrub by the road, or from apples, weeds, sawdust–almost anything. There is fuel in every bit of vegetable matter that can be fermented." However, these alternative bioenergy assets are exceptionally ecological agreeable, they must be assessed on case-to-case reason for their pros & cons and explicit applications. This chapter analyse critically the impacts of biofuels on both ecological and economic aspects in various forms.

2 Reasons for Introducing Biofuels Worldwide

Biofuels have been a piece of the energy for quite a long time. Nonetheless, in the course of recent years, conversation and activity on biofuels has expanded with ascends in raw petroleum costs. In any case, notwithstanding costs, there are lot of justifications available for the governments to show interest in biofuels in any event, when sponsorships are expected to make them monetarily suitable are mentioned below [2].

(a) **GHG reduction:** Much researches demonstrate that the utilization of bioenergy fuels reduces GHG emission compared with fossil fuels (IEA, 2004) though the extent of reduction of GHG depends on crop and production technology. Some studies indicate that biofuel production generates more GHG than it saves in burning.

(b) **Carbon-free:** An international treaty, "Kyoto Protocol" in 1992 said that CO_2 emission from the burning of biofuel can be offset by ingested CO_2 while development of the plant for the production of biofuels. We speculate that using biofuel is a countermeasure against global warning and unnatural weather change. This is the main justification behind presenting biofuels on fundamental level.

(c) **Renewable Energy**: Biofuel is believed to be a sustainable power since it is derived from plants and isn't depleted except if impeding the development of the plants. Since biofuel is renewable which is inexhaustible, it has gained most widespread acceptance all over the world.

(d) **Reduction of Air contamination**: As compared with fossils fuels, burning of biofuels causes reduction in the CO_2, CO, N_2O particulate matter, and many other harmful toxic gases, which helps in controlling air pollution. Even when biofuels like bioethanol, biodiesel mixed with petroleum fuels it makes combustion complete due to the presence of oxygen in biofuels. By this way it also controls air pollution.

(e) **Ensures Energy Security:** The instability of world oil prices, unbalanced worldwide circulation of oil supplies, and a substantial reliance on importing oil fuels forces the oil importing nations helpless against supply. These could be reduced by the creation of biofuel from plants filled in a country, which might add to energy security in that country.

(f) **Rural Areas development**: The production of biofuels creates new jobs opportunities in the agricultural sector in many rural areas of developing nations which helps in expanding the economy of the farmers and that of the country by higher the cost of the exporting commodity products.

(g) **Trade balance**: Many countries which imports crude oil to satisfy the energy need spend a huge amount of their reserve income in importing this oil. This can be reduced by producing biofuels from the agricultural products produced within the country.

3 Generations of Biofuels

Biofuels may be ordered based on the source from which it is derived. An interesting fact is that the structure of the biofuel doesn't change between generations, rather the source (feedstock) from which the fuel is derived changes. The pros and cons of the 1st generation biofuels lead to the improvement and development of consecutive generations of biofuels.

(a) **1st generation biofuels**:
Any feedstock which is directly produced from food crops which are grown in agricultural land that can be taken as food by humans are said to be the 1st generation of biofuels. Some of the well-known crops used as 1st generation biofuel feed stock are corn, wheat, sugarcane, rapeseed, etc., that provides sugar, fats, starch, etc. for producing biofuels. It include bioethanol and biodiesel. Ethanol is usually produced from the glucose fermentation process whereas the biodiesel from by transesterification process.

(b) **2nd generation biofuels**:
These are advanced biofuels. The feedstock for the production of this category of biofuels are usually of non-food products. It is only if it has effectively satisfied the need for the demand of food and is presently not good for human consumption. It is advantages that it does not require agricultural land for the growth of the feedstock, it could be developed on marginal land and can be utilized.

(c) **3rd generation biofuels**:
The 3rd generation biofuel is derived from the algae biomass. It has a distinct growth yield compared to 2nd generation biofuels. In the beginning, algae was also placed in the 2nd generation biofuels. But later, due to the potential of algae in providing a huge yield it was separated from second and placed as third generation biofuels. Coming to the potential of algae to produce fuel, no feedstock can match it in terms of quantity and diversity of fuels. It is to be noticed that the production of algae biomass don't need cultivation land for its growth and it is also not a food crop.

(d) **4th generation biofuels**:
This generation of biofuels combines genetically engineered algae with genomically produced micro-organism to effectively generate biofuels. They are grown using non-arable land, similar to that of the 3rd generation biofuels. Researches in this 4th generation biofuels is now at its primary stage. Its research has started from 2006 and still goes on and many significant results has not yet been published (Fig. 1).

4 Ecological Impacts of Biofuels

The effect of biofuels creation and its use on environment and climate is profound and broad. The increase of agricultural production systems and the transformation of existing croplands for biofuel feedstocks production will have a drastic impact on our environment beyond their impacts on GHG emissions.

4.1 Impacts on the Atmosphere

The major cause of air pollution and climatic changes occurs due to the emission of Green House Gas (GHG). Figure 2 explicitly shows that the emission of various GHG has been drastically raised during the past two decades due to the industrial revolution which intakes large amount of energy especially from fossil fuels. Numerous law-makers accepted that the fossil fuels can be replaced with biofuels will have a huge positive environmental and climatic effects by emitting less amount of GHG which add to a dangerous atmospheric deviation.

Biofuel feedstock can diminish or counterbalance GHG liberation by directly eliminating carbon dioxide from the atmosphere and it can be recycled and reused by the feedstock crops. Despite these advantages of biofuels, researches shows different GHG balance for different biofuels when compared to fossil fuels.

The life cycle assessment of the GHG emission from the sources for the production of bioethanol are as follows [3],

- Production stage (Agriculture products)
- Transportation stage (Supply)
- Bioenergy producing stage (Bioethanol)
- distribution stage (Usage)

LCA of CO_2 emissions of domestic ethanol production = CO_2 emission from the material resource end of the energy crops + CO_2 emission from the ethanol production end−apportioned CO_2 emission from byproduct use or CO_2 deduction of byproducts as the energy input for the production process [4].

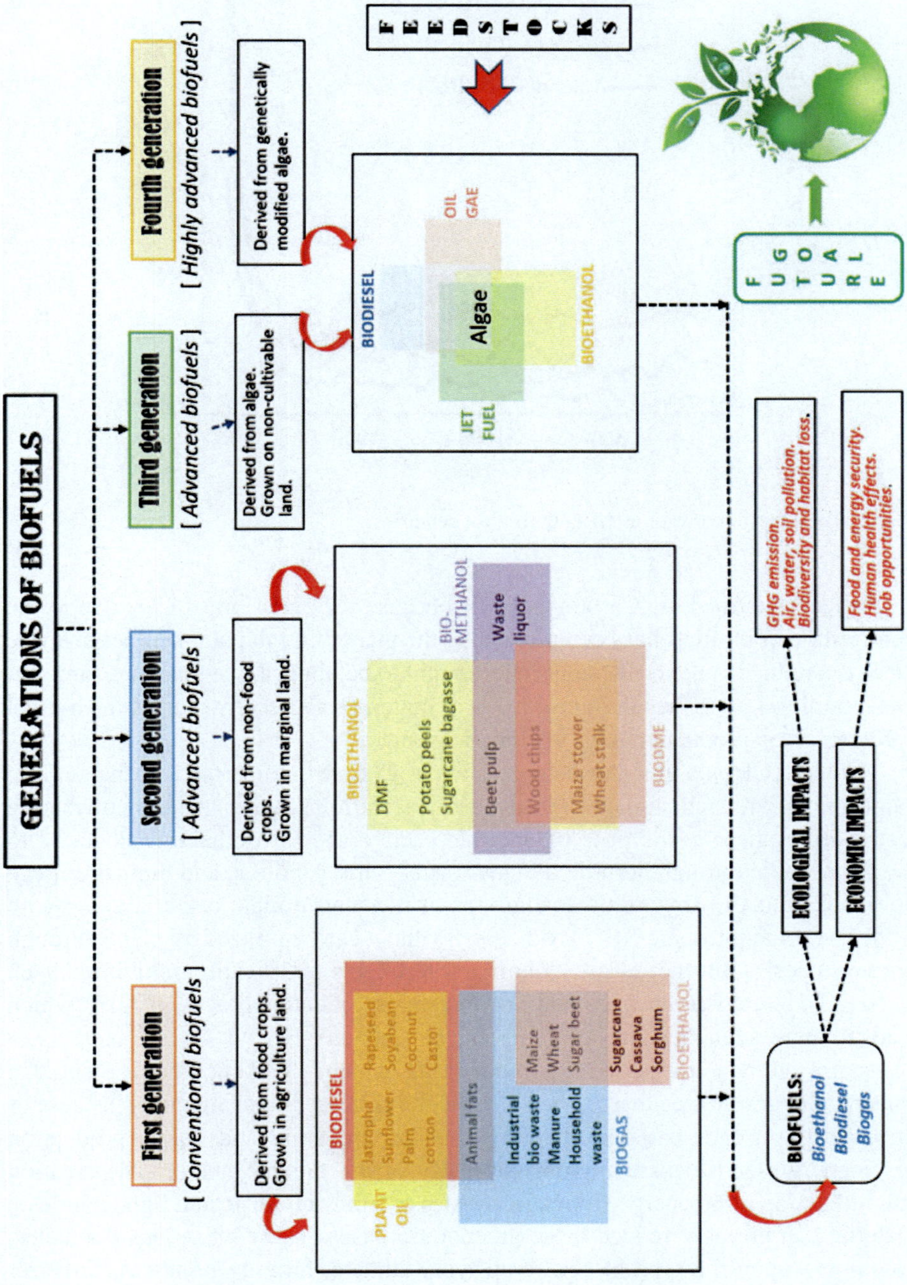

Fig. 1 Generations of biofuels, its feedstock and its impacts

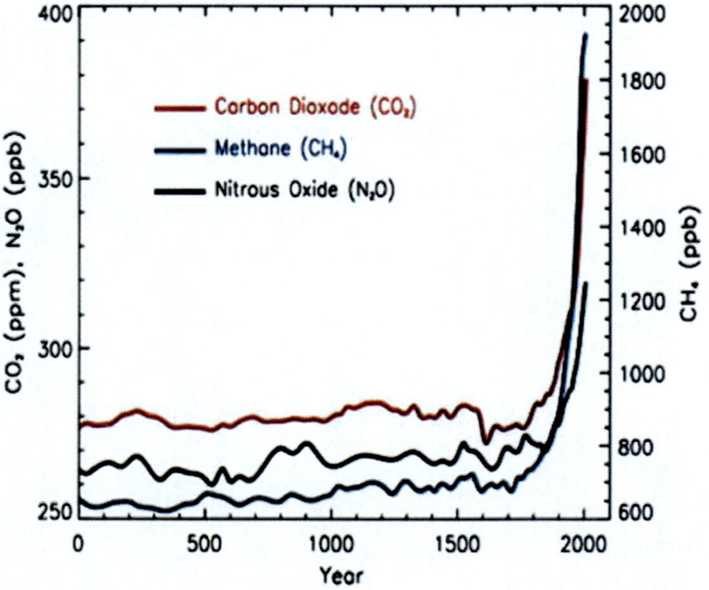

Fig. 2 The emission of various GHG in the past centuries

Biofuel is termed as 'Carbon–neutral' because the emission of CO_2 due to the combustion of biofuels has been utilized by the plant through photosynthesis, hence there is a reduce in net GHG emission. It ought to be noticed that there are some net GHG outflows, as the result of utilization of materials and energy during the biofuel feedstock crop production adds to climatic change.

Methane (CH_4), a greenhouse gas which is 15 times more potential than CO_2in capturing infrared radiation and results in global warming. It is expanding in fixation in the atmosphere at the pace of one rate each year [5]. Methane is liberated to the atmosphere through burning of agro-waste, cattle, landfills, and biomass. It can be oxidized to CO_2 by the diverse groups of methano trophic bacteria along with some chemo-autotrophic NH^{4+} oxidizers. So that it can be utilized by plants through photosynthesis. But the release of nitrogen fertilizers, nitrification inhibitors (such as NH_4, NH_3) into the environment can prevent the oxidation process of CH_4, which leads to increase in its concentration in the atmosphere.

Another greenhouse gas, Nitrous oxide (N_2O) which is 300 times more potential in causing global warming than carbon dioxide. It is released fundamentally because of the usage of nitrogen-based fertilizers used in the cultivating lands for the production of bioenergy feedstocks and from the decay of the organic matter. The creation and utilization of bioenergy from agro-wastes or from cellulosic and lignocellulosic material can possibly reduce those greenhouse gases. These emissions fluctuates, contingent upon the type of soil, crop type, climate, farming practices, fertilizer types and application recurrence. The N_2O emission reduction potential of different types of biofuels based on the different land-use change are tabulated below (Table 1).

Table 1 The N_2O emission reduction potential of biofuels [5]

Biofuels	Feedstock	GHG emission due to fossil fuels (%)	Reduced GHG emission by use of biofuels (%)	Region
Bioethanol	Sugarcane	103	62	All regions
Biodiesel	Palm oil	75	39	All regions
Bioethanol	Sugar beet	58	17	East Europe
Bioethanol	Sugar beet	58	21	West Europe
Bioethanol	Wheat	90	6	North America
Bioethanol	Wheat	75	5	EU25
Bioethanol	Wheat	107	53	Oceania
Bioethanol	Corn	38	2	North America
Bioethanol	Corn	36	11	EU25
Biodiesel	Rapeseed	72	13	North America
Biodiesel	Rapeseed	76	9	EU25
Biodiesel	Rapeseed	80	75	East Europe
Biodiesel	Soyabean	111	30	North America
Biodiesel	Soyabean	87	44	South America

Let us consider an example of Indian scenario on biofuel production. The burning of rice straw left in the agricultural land to clean it for the following yield produces a ton of destructive vaporous into the climate that prompts the air contamination. Consistently, around 15 metric tons of rice straws is singed alone in Punjab. Consuming the yield build-ups acquired from rice and wheat cultivating system of Punjab, Haryana and Uttar Pradesh lead to the outflow of ozone depleting substances which at last prompts the climatic change. The burning of this feedstock creates emission of 70% CO_2, 7% CO, 0.66% CH_4, 2.09% N_2O and various harmful gases such as hydrocarbons, NO_2, NO_3, SO_2, etc., and many other particulate matter are emitted during this open burning. Utilizing lingering rice straw to produce bioethanol is a practical method of waste use. The usage of agricultural waste for power production is a green drive for the production of energy need. Though, there is an outflow of GHG during the conversion of feedstock to biofuels, it is extremely less when contrasted with petroleum products.

Many National and International policies on climatic changes says that there are several ways to get down the Greenhouse gases emissions, but all of these has to be implemented and followed forcefully to accomplish an eminent climate and environment. Specifically, every area of the economy should take critical measuses to profound carbon stops within a limited timescale before 2050.

As to conclude, form the number of review studies says that there is a huge decrease of GHG emissions into the atmosphere when compared to fossil fuels. A study analysed the carbon–nitrogen footprints for both 1st and 2nd generation biofuels versus fossil fuels shows that the C-footprints of 1st generation biofuels is huge and it is undoubtfully reduced by utilizing second-generation fuels. Plainly the usage of biofuels will supplant fossil fuels which plays a vital part in reducing CO_2 emissions to the atmosphere [6].

4.2 Impacts on Water Resources

The feedstock for producing biofuels are the rural harvests. More than 70% of the world's freshwater available is utilized for agricultural purposes, so water shortage might end up being as a significant factor to be considered for the producing biofuel feedstock plant crops. Water resources for farming are turning out to be more and more scant in many nations on account of the expanded rivalry with domestic or industrial usage. In addition to this, climatic changes will further create stress on already existing water demand.

Many of the biofuel producing feedstock crops such as sugarcane, maize, wheat, corn, etc., have relatively high water requirements for commercial yield. Further, the production of biofuels for these feedstocks will also utilize enormous amount of water. In this way, the creation of biofuel yields will have a direct and harmful influence on the quality and quantity of the water resources. Biofuels production results in organically debased wastewater, if it discharged untreated, would build eutrophication on the surface of waterbodies. Since bioethanol and biodiesel are biodegradable, as it contains more amount of oxygen, it reduces the harmful impacts on soil and water resources. For example, In Brazil, even though the sugarcane for bioethanol is grown primarily under rainfed conditions, it requires 4–8 L of water, just for the production of 1 L of bioethanol along with 13 L of waste effluent. Considering all these into account, the water contamination is related with all the stages from usage of manure and agrochemicals, sugar-cane washing, surface runoff, so on to ethanol production process. Still water stays as a major issue for bio-ethanol production [7, 8].

The fertilizer used in the fields of biofuel producing crops increases the nitrogen and phosphorus content of the soil, leads to eutrophication, which causes many harmful impacts on aquatic ecosystems by contaminating the water, reducing species richness of aquatic ecosystem along with the rapid growth of harmful algae [9]. It also leads to Oxygen demand and interfere with utilization of the water for fisheries, industry, agriculture, and drinking purposes and causes death and decaying of aquatic plants and fishes, which creates an imbalance in the aquatic ecosystem. As the contamination from the fertilizers and usage of pesticides for the biofuel producing crops creates a huge impact on terrestrial as well as the aquatic ecosystem.

4.3 *Impacts on Soil Resources*

The feedstocks for producing biofuels are the farming crops. Due to the change in usage of agricultural land and increase in crop production for the feedstock will both unfavourably affect soil.

(a) **Impacts on soil chemical properties:** The removal of plant residues by preharvest burning will cause air pollution. In other case, if we use the plant residue as the feedstock for the production of 2nd generation biofuels in-order to reduce air pollution, will lead to decrease in soil nutrient content as the soil natural matter assumes the significant part in the useful limit of many tropical soils is the significant concern.

A study conducted on changes in soil properties under consistent sugarcane production was examined in two ways. In any case, soil properties are checked over a period of time at a similar region and in the subsequent methodology, soils under nearby unique land-use frameworks are tested simultaneously. The results shows that the continuous plantation of sugarcane leads to the considerable decrease in the soil pH, soil organic contents like organic C, available P, nitrogen content, exchangeable ions in the soil, which will lead to the discrepancy in soil fertility, crop yield under continuous cultivated land [10]. The solution would be the crop rotation method, which will continuously regain the soil nutrient content. Therefore, crop rotations, conservation tillage and other management practice can reduce the adverse impact on the environment.

(b) **Impacts on Soil Physical Properties:** The plantation of sugarcane requires large machinery, which are utilized for land preparation, harvest, applications of fertilizers, and pesticides. These instruments affect the physical properties of soil like bulk density, aeration and porosity. Usually, water inlet is reduced by increase in the topsoil bulk density. Usually, an uncultivated land have a density of 1.40 Mg m^3. There is an increase in the bulk density of soil of 0.15–0.18 Mg m^3 in the topsoil of the cultivated land compared to uncultivated land soil erosion may also be of significant concern [11, 12]. Many crop residues such as rice straws wheat straws, corn straws and many other little grains are left over the land after harvest to reduce soil erosion. But these forms the basic feedstock materials for the production of 2nd gen-biofuels. If it is removed from the land for the production of biofuel will lead to the reduction of surface topsoil and increases soil erosion [13]. To decrease soil erosion by half of that happening for an exposed soil surface, nearly 30%of the soil surface should be covered by crop residues followed by plantation of new crops [14]. The procedures, for example, cover cropping and crop rotation are need to be followed strictly to reduce soil erosion.

(c) **Impacts on soil biological properties-** The changes in soils physicochemical properties due to continuous sugarcane harvest in the same field will affect the biological properties of the soils. It is found the large groups of the microbial community responded to particular environmental attributes. For example,

Bacteria like actinomycetes will generally in contact with soil organic matter, while other organisms like fungus, protozoa are related with the heavy metals of soil [15]. So, the change in the soil physiochemical properties directly affects the soil microbial community.

The macrofauna (like earthworms) are the significant parts of the agro-framework of soil, as they add to the physicochemical properties of the soil. But some mechanical methods (like tillage practice) used for crop cultivation causes direct damage to the earthworm and reduces the earthworm population in soil will lead to the deficiency in soil organic matter, soil water storing capacity, increasing the soil temperature [16]. Researches shows that the earthworm population are higher in the pastures than the conventionally tilled cropping system.

4.4 Impacts on Biodiversity and Habitat Loss

Biodiversity can be defined as the variability among variety of land, species and organism but biofuel feedstocks depends on land, as of now homogeneous should as much as possible. There gives off an impression of being an intrinsic clash. The expansion in the bioenergy feedstock creation drives the land use change which essentially impacts encompassing environments, improving or smothering wild and rural biodiversity.

Some of the positive impacts on biodiversity is that it leads to the restoration of degraded lands for cultivation of bioenergy feedstocks, which can improve biodiversity and fill gaps between the remaining parts of natural habitat. Even though it has some positive impacts, significant number of its effects will be negative. For instance, when regular scenes are changed over into energy crop estates, it will prompt draining of peat lands [17].

In two different ways, biofuel feedstock production leads to the loss of biodiversity. The primary pathway for biodiversity loss is through loss of habitat. The most extreme type of habitat loss happens when a diverse groups of plants, creatures and micro-organism is supplanted by monoculture. The need and demand for the biofuels lead to the expanding of tropical and plantation forest to biofuels feedstock crop production. All these add to the loss of biodiversity and habitat of the wildlife species.

The currently available land in this world is designated to different land utilities is given by

$$Total\,land\,area = Agricultural\,land + Conservation\,land + Urban \tag{1}$$

In case of biofuels, Eq. (1) can be altered as land required for biofuel production need to be utilized from the currently available areas,

$$Total\,land\,area = Food\ production + Biofuel\ production + Conservation + Urban \quad (2)$$

Land for biofuel production essentially consummate with the total land, so it will compete directly with conservation land and indirect competition occurs with agricultural land [18]. There exist a criticism encompassing the likelihood that regions with high natural worth with high biodiversity will be cleared for sugarcane production may destroy these rich environment and biodiversity [19].

Another major contributor for the loss of biodiversity is the use of agrochemicals, mainly of nitrogen, phosphorus, as the input for intensive energy crop production, which makes the way for chemical fertilizers from agricultural land to other habitats. Extension of biofuel creation might involve expanded utilization of chemical applications to augment yields. The results of an empirical investigation on inorganic fertilizers use and biodiversity risks, indicates that mineral fertilizers use is altogether identified with biodiversity hazards, which implies that increase in the use of inorganic manure will prompt expansion in degradation of biodiversity. Even though we can't track down a steady example of effects on biodiversity hazard, at any rate, the confirmations supporting the speculation that the power of inorganic fertilizer will have expanding hazard on biodiversity [20]. The utilization of inorganic fertilizers for several decades has differential impacts for various groups of organisms, for example, birds, animals, earthworm, microorganisms, etc. The application of organic fertilizers and biotechnology will fundamentally add to the sustainability of agriculture. The use of Burnt Municipal Solid Waste (BMSW) for producing biofuels especially bioethanol purpose can also avoid this problem. Since BMSW does not require any fertilizer or agrichemicals for biofuels biofuel production, it is eco-friendly and also has economic and social benefits [21].

5 Economic Impacts of Biofuels

The public authority's purposes behind supporting biofuels, at first is simply monetary, along with energy security, GHG emission, worldwide climatic change, and rural development. Supporting the economic development along with agriculture or through agriculture is an important objective of current cultures to work on the riches and prosperity of the residents of the country.

5.1 Energy Security

The world burns-through around 400 EJ of energy each year yet produces what could be compared to around 100 EJ of energy generally from unused harvest crop as it could deliver 180 extra EJ from energy-devoted plants and trees. The oil shocks of 1973 and 1979 caused oil costs to take off to US$40 a barrel, pushing Brazil's yearly use for oil imports to more than US$10 billion and causing a worldwide downturn. To cover these huge bills and develop to develop more domestic energy, Brazil imported more from abroad. In the mid of 1980s, considerable expansion in financing costs overall constrained Brazil, alongside other Latin American nations, to carry out severe monetary changes that led to negative economic growth and rapid inflation. Since 1975 bioethanol has uprooted in excess of 280 billion litres of fuel and saved more than US$65 billion. At the point when the expense of overhauling the obligation that such imports would have required is incorporated, the expense reserve funds ascend to more than US$100 billion [22, 23].

5.2 Employment and Job Opportunities

The development of biofuel industries will bring openings for the socio-economic development particularly in developing countries. Investment in new biofuel sector affects other economic sector which will bring about expansion in economic activity. The expansion of market opportunity for biofuel feedstock crops and their increasing prices on global market will increase employment and income generation because it offers jobs for both in agricultural and processing sectors [24].

For developing nations with countless needy individuals dependent on agriculture, the main goal of the nation is the successful utilization of farming residues for energy production. This is a best choice for creating extra income to poor country communities with less impacts on poor people [25].

An efficient bioenergy sector, which begins with biofuels feedstock production will helps in rural development as it is characterized by the labour intensive action. If the feedstock produced is utilized for the biofuel production will create a demand for the agricultural food products which consequently lead to increase jobs and wages in farming sectors. This reality accentuates the connection between the biofuels sector and rural development [26].

5.3 Health Issues

The World Health Organization (WHO) said that "**the vegetation fuels produce adverse respiratory health effects on human beings**."

Traditional bioenergy utilized by the poor, depend heavily on biomass which is mostly composed of burning wood and agricultural waste will have extreme adverse consequences on women and children as they are exposed to burning ashes and smokes with in a bounded space. This exposure will have extreme health issues such as coughing, wheezing, asthma and other respiratory sickness [26] (Fig. 3).

The biofuels have huge impacts on health of the human well-being in three ways.

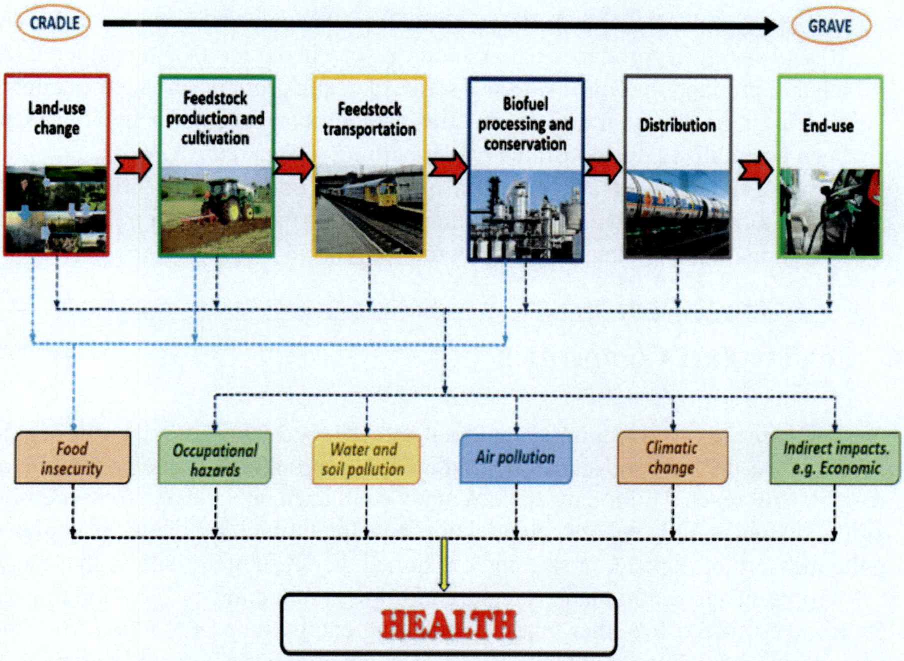

Fig. 3 Schematic representing the LCA of biofuel production associated with health issues

(a) During the production of biofuels, the chemicals and biochemicals are used, which are hazardous to human health and surroundings. Some of the chemicals used in the production of biofuels are listed below,

i. Sulfuric acid (H_2SO_4) is used during fermentation of ethanal to adjust the pH of the mixture.
ii. Sodium hydroxide (NaOH) and potassium hydroxide (KOH) plays a catalytic role for the biodiesel production.
iii. Sodium hydroxide (NaOH) of used in sterilization and cleaning.
iv. Ammonia (NH_3) is used as nitrogen source.
v. Hydrogen sulphide (H_2S) is released during the production of biogas, etc.

These chemicals have direct effect on human health such as skin irritation, respiratory disease, etc. [27].

(b) The preparation and utilization of biofuel products releases different air toxins, including CO, NO_2, NO_3, HC, suspended particulate matter, and unpredictable natural mixtures. Some of the unburned toxic components present in air leads to the formation of fog and smog [28]. In the cities surrounded by sugarcane plantation, exposed to huge amount of air pollution, because there is a continuous burning for at least 6 months, as it generates large quantities of particles and toxic substances which directly affects the entire human race. These pollutants are play a major role in increasing morbidity especially cardiovascular and respiratory diseases and even certain type of cancers.

For example, Brazil is the largest producer of sugarcane in the world. About 70% of sugarcane produced in its country is scorched prior to being gathered for ethanol production. This results in a series of health problem among the country people. In a city of Araraquara in Brazil, more than 1,92,000 inhabitants had been hospitalized due to asthma by the burning of sugarcane [29].

(c) Besides air pollution and industrial processing pollution, the production of biofuels could directly influence human wellbeing through water pollution, soil contamination and many other work related risks.

6 Food to Fuel Competition

If we compare the world requirement of food with the food supply, it shows the degree of security. As per the Food and Agricultural Organization (FAO), total population of the world will reach 9 billion by 2050. A study conducted on, land region needed for worldwide food supply and its potential of producing biofuels which was calculated by the measure of population size and consumed diet in future. It shows that about 55% of present agricultural land is required for food production by 2050 and the rest 45% can be utilized for other purpose including energy crop production. But, the demand biomass is higher, which lead to more utilization of potential agricultural land area. In this way, there will be a contest for a similar land region for food and biofuel, which will increase the food prices. The poor in both urban and rural areas will suffer due to increase in food prices. This may threaten global food security [30]. Figure 4 indicates the increase in the total share of food prices due to world biofuel production [31].

The poor need to spend a huge portion of their pay for food, of more than half of their income in many cases. Also, numerous developed countries import food to fulfil the domestic need of the country. The country with a high populace of food start to trade grains, to exploit high fare costs and the nations which can't import food because of the high grain costs, experiences lack of healthy sustenance and increase in hunger [32]. So, it is necessary to keep biofuels from demolishing the food-costs crisis.

Lack of investments in agricultural research in recent years was the major contribution to slower the rate of world agricultural production when it demands more. Even though biofuels are just one of the few hotspots at expansion in food costs will still continue to create a huge impact on increase in food prices. So, it necessary to take into account the poor people with a well-developed plan to help them to access the food [33].

For example: The U.S. livestock sector was antagonistically affected by the sharp expansion in feed costs particularly in corn grains in 2008 [34, 35]. It was criticized for being the major element for raising the cost of corn and different grains and food varieties. With their biggest expense class expanding drastically, net returns for producers have been feeling the squeeze.

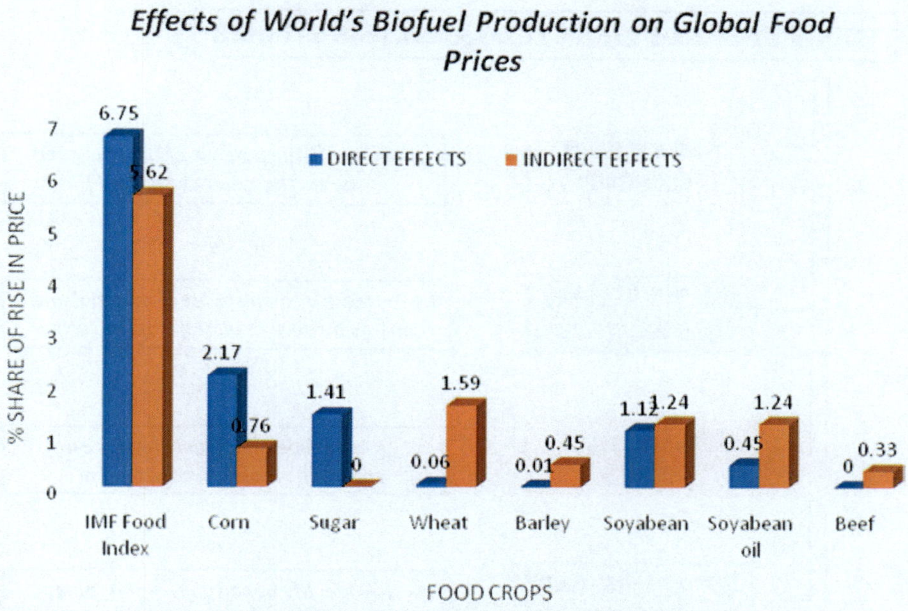

Fig. 4 Effects of World's Biofuel Production on Global Food Prices

During world food cost emergency (2007–08), Report by a World Bank named, " A Note on Rising Food Prices" said that biofuel production was the major reason for more than 70% rise in the costs of food wares. Then again, another World Bank paper in July 2010 tracked down that the effect of biofuels on food costs during the 2008 value blast was "significantly less than at first suspected" and "index fund activity" assumed a significant part in the value blast. In this way, the discussion on the effect of biofuels on food costs is as yet unsettled. Different components like stock market deliberation, environment factors, and increasing expense for petrol could likewise play played significant parts in driving up the food costs in 2008 [36]. Thus, it is important to take some Mandatory measures in bringing down the food costs, which will keep on driving the future strength of the agricultural sector and will assume a biggest part in deciding the food security and human prosperity of the world's poor and more weak population.

7 Biofuels–Lifecycle Assessment (LCA)

Advancement in different techniques for the better understanding and addressing the effect of biofuels, one of the best improvements for this reason for existing is Life Cycle Analysis (LCA) strategy. It is an incredible asset to assess economic proficiency and effects on nature or its interaction.

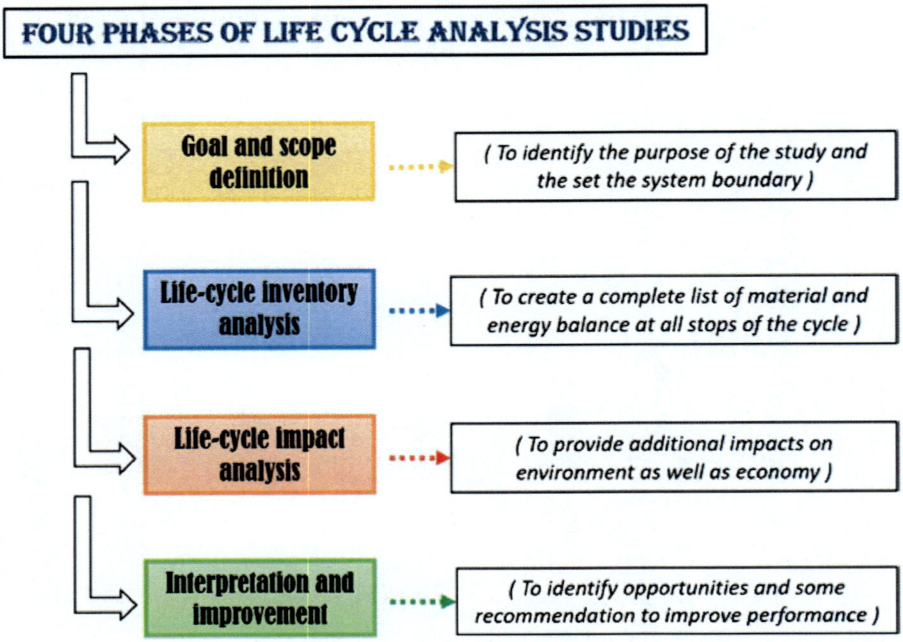

Fig. 5 Four phases of LCA studies

As indicated by NREL, Life-cycle assessment (LCA) is "**a standardized technique that tracks all material, energy, and pollutant flows of a system from raw material extraction, manufacturing, transport, and construction to operation and end-of-life disposal**."

LCA tends to the ecological effects alongside financial and social perspectives from Cradle-to-grave. A Cradle-to-grave LCA investigation considers the whole life-pattern of a product from its creation (Cradle) to its last removal (Grave). This study consist of four stages which are represented below (Fig. 5).

The LCA method of analysis is utilized as a standard method for analysing the life-cycle of GHG emission from biofuels. For example, a study conducted on biofuel LCA of GHG emission, in which they consider four different methodologies for assessing, and found that their procedure encourages the target of reducing GHG emission from biofuels production to its final usage [37].

An LCA study of biofuels has been conducted to analyse the impacts of water use in reference with fossil fuels. It shows that, the overall water consumption use for biofuel production along with the background process such as fertilizer production, etc., is equivalent to that of the fossil reference when no irrigation occurs and it dominates when the agricultural irrigation is higher [38]. Beside the reduction of GHG emission from biofuel as compared with fossils, it leads to impact the ecosystem.

8 Sustainable Agriculture and Sustainability of Biofuels

The expression "Sustainable agriculture" portrays a financially feasible, ecologically protected and socially acknowledged food, fiber or fuel creation framework. The idea of sustainability was first introduced by the UN Conference on Environment and Development held in Rio de Janeiro.

Developing practical food system, adds to the supportability of the human populace. For instance, probably the most ideal approach to relieve environmental change is to make reasonable food frameworks dependent on sustainable agriculture. It gives a possible answer for empower rural frameworks to take care of a developing population inside the changing natural conditions [39].

Environmental sustainability can be seen as adjusting the three mainstays of financial and social improvement with ecological protection. For a practical turn of events, the improvement of economy and climate ought to go together as both are fundamental for the advancement of people. By and by, the primary issue of world is the way to diminish outflow of ozone depleting substances for relief of environmental change and to ensure the maintainable development of economy. This can be accomplished by advancing new sustainable wellsprings of energy like wind energy, solar oriented energy, biomass based energy, etc.

A few researchers have created key standard of sustainability,

i. multitudinous approaches on ecological, economic and social viewpoints.
ii. Extremely adaptable and dynamic techniques of evaluation and execution of complex frameworks (like diverse yield societies, crop pivots, ecological conditions).
iii. A fundamental examination that assesses single factor as well as perplexing capacities and cycles.
iv. Transparent ideas proposing clear proposals for execution.

Developing requirement for biofuels and bioenergy creation made us to look for a standard that tends to attain the goals of sustainability. This requires a specific production framework which is naturally, socially and financially maintainable. Moreover, this needs to add to decrease of GHG outflow, creates no adverse consequences just as adding to positive social results (Fig. 6).

The sustainability goals aimed to ensure

i. To minimize GHG emission.
ii. To prevent high biodiversity land.
iii. To provide environmental requirements for agriculture.
iv. To prevent the land, water, air resources for future needs.
v. To protect forest by significant human activities.
vi. To conserve the ecosystem.
vii. To develop human resources.
viii. To protect nature and nature areas.

The need of energy never reaches a conclusion so; the test is to acquire power source adequate to present for our energy needs. Furthermore, this energy source

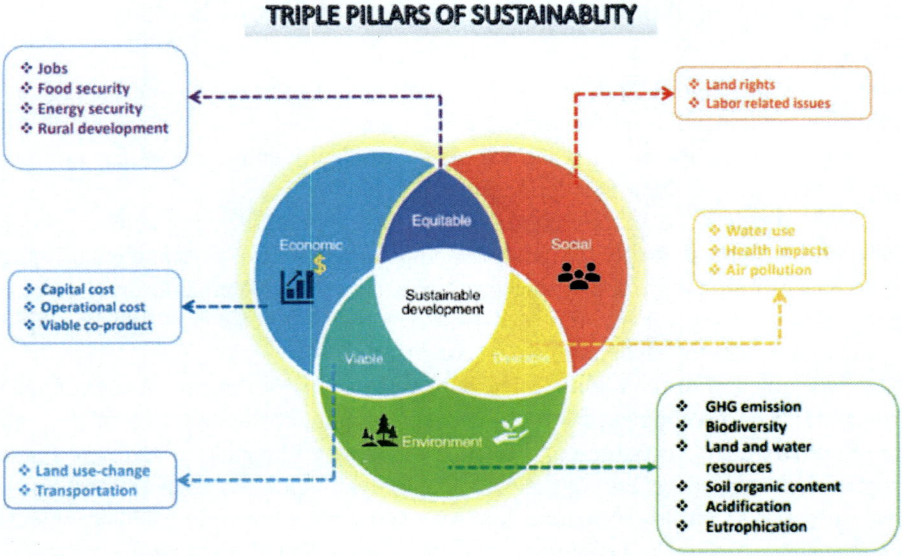

Fig. 6 Sustainable development. (*Source* Modified from Reality Check: A critical look at MIT's fuel consumption research)

should be trustworthy, inexhaustible, repeating and non-adding to environmental change. The inexhaustible asset of energy actually having limits and their technical advancement have some positive and negative impacts. 1st generation biofuels show a development toward cleaner and environmentally friendly power, it actually lingers behind as they show up from biomass which is a food item and due to its monetary components. A few advantages are likewise given from second generation type′s biofuels as it comes from basically non-food biomass and from district squanders however the biomass requires versatile advances and contends with food crops over arable grounds in a few pieces of the earth. The third era biofuels relate to the most assumption, there are still a few difficulties in introducing them financially more feasible. Notwithstanding, loads of exploration actually wants to be done to reduce down assembling expenses and make this sort of fuel creation industrially, reasonably practicable.

9 Conclusion

The need of energy never comes to an end. The bioenergy apparently is, to an ever increasing extent, as promising and generally undiscovered sustainable power asset, and its potential environment and financial advantages are turning out to be clear as technological improvements continue to emerge.

Yet, around the world, there is just one organization, Logen Corporation, Canada, produce bio-ethanol at business scale utilizing wheat straw and corn stover. In India, in spite of abundant accessibility of biomass, there is no commercial bioethanol production plant from lignocelluloses.

As to presume that in spite of the fact that there are significant and ethically pertinent contrasts between different methods of rural creation, given the present and extended size of the human populace, eating grain-took care of creatures and changing food over to fuel are hard to morally legitimize. Till now, biofuels have primarily been assessed as for the decrease they acquire GHG outflow, while the potential expanding in water utilization use and land use change has gotten just a minor consideration. In fact, water use, notwithstanding land usage, could turn into a significant concern, counterbalancing all advantages of biofuels. The eventual fate of biofuels is a fervently discussed point. Perspectives on scientists shift on pretty much every issue except one thing that biofuels are digging in for the long haul and struggle can emerge from a few fronts. Despite the fact that GHG alleviation from first generation biofuels shifts to the consecutive generations of biofuels hold more freedoms for opening the capability of biofuels and assist with accomplishing environment relief targets however they are not yet financially suitable and the innovation are yet being created. Bioenergy will be a ‘significant part of energy mix’ later on.

References

1. Chandel AK, Chan ES, RavinderRudravaram ML, Narasu LV, Rao and PogakuRavindra, (2013) Economics and environmental impact of bioethanol production technologies: an appraisal. Afr J Environ Econ Manag 1(5):126–143
2. Dufey A (2006) Biofuels production, trade and sustainable development: emerging issues. International Institute for Environment and Development, London
3. Quirin M, Gärtner S, Pehnt M, Reinhardt GA (2004) CO_2 mitigation through biofuels in the transport sector. Status and perspectives. Institute for Energy and Environmental Research Heidelberg, Germany, p 55
4. Meihui S, Chiahui H, Wenyi L, Chunto T, Huusheng L (2014) A multi-years analysis of the energy balance, green gas emissions, and production costs of first and second generation bioethanol. Int J Green Energy 12(2):168–184
5. Smeets EMW, Bouwman LF, Stehfest E, Van Vuuren DP, Posthuma A (2009) Contribution of N_2O to the greenhouse gas balance of first-generation biofuels. Glob Change Biol 15(1):1–23
6. Čuček L, Klemeš JJ, Kravanja Z (2012) Carbon and nitrogen trade-offs in biomass energy production. Clean Technol Environ Policy 14(3):389–397
7. Moreira JR (2006) Bioenergy and agriculture, promises and challenges: Brazil's experience with bioenergy. Vision 2020, Focus 14, Brief 8 of 12. Washington, DC, International Food Policy Research Institute.
8. Moreira JR (2007) Water use and impacts due ethanol production in Brazil. Paper presented at the International Conference on Linkages between Energy and Water Management for Agriculture in Developing Countries, ICRISAT Campus, Hyderabad, India, 29–30 January 2007. São Paulo, Brazil, National Reference Center on Biomass, Institute of Electrotechnology and Energy, University of São Paulo
9. Carpenter M, Gärtner S, Pehnt M, Reinhardt GA (2004) CO_2 mitigation through biofuels in the transport sector. Status and perspectives. Institute for Energy and Environmental Research Heidelberg, Germany, p 55

10. Hartemink AE (2008) Chapter 3 Sugarcane for bioethanol. Advances in agronomy, pp. 125–182
11. Maclean NR (1975) Long-term effects of sugar cane production on some physical and chemical properties of soils in the Goondi mill area. Proc Qld Soc Sugarcane Technol 42:123–126
12. Sanchez ST, Woods J, Akhurst M, Brander M, O'Hare M, Dawson TP, Malpas R (2012) Accounting for indirect land-use change in the life cycle assessment of biofuel supply chains. J R Soc Interface 9(71):1105–1119
13. van Ouwerkerk ENJ, Anex RP, Richard TL (2007) Web-based and database driven biomass planning tool. In: Proceedings of farming systems design, pp 10–12
14. Laflen JM, Colvin TS (1981) Effect of crop residue on soil loss from continuous row cropping. Trans Am Soc Agric Eng 24(3):605–609
15. Liang C, da Jesus EC, Duncan DS, Jackson RD, Tiedje JM, Balser TC (2012) Soil microbial communities under model biofuel cropping systems in southern Wisconsin, USA: impact of crop species and soil properties. Appl Soil Ecol 54:24–31
16. Nunes DH, Pasini A, Benito NP, Brown GG (2006) Earthworm diversity in four land use systems in the region of Jaguapita, Parana State, Brazil. Caribb J Sci 42:331–338
17. CBD (Convention on Biological Diversity) (2008) The potential impact of biofuels on biodiversity. Note by the Executive Secretary for the Conference of the Parties to the Convention on Biological Diversity, 19–30 May 2008, Bonn, Germany (draft, 7 February 2008)
18. Sala OE, Sax D, Leslie H (2009) Biodiversity consequences of increased biofuel production, Department of Ecology and Evolutionary Biology, Brown University, Providence, RI, 02912, USA
19. Martinelli LA, Filoso S (2008) Expansion of sugarcane ethanol production in brazil: environmental and social challenges. Ecol Appl 18(4):885–898
20. Mozumder P, Berrens RP (2007) Inorganic fertilizer use and biodiversity risk: an empirical investigation. Ecol Econ 62(3–4):538–543
21. Li A, Khraisheh M (2010) Municipal solid waste used as bioethanol sources and its related environmental impacts. In: Proceedings of the annual international conference on soils, sediments, water and energy, vol 13, Article 12
22. Hazell P, Pachauri RK (2006) Bioenergy and agriculture, promises and challenges: Brazil's experience with bioenergy. Vision 2020, Focus 14, Brief 1 of 12, International Food Policy Research Institute, Washington, DC
23. Hess P, Johnston M, Brown-Steiner B, Holloway T, de Andrade JB, Artaxo P (2009) Air Quality issues associated with biofuel production and use
24. Silalertruksa T, Gheewala SH, Hünecke K, Fritsche UR (2012) Biofuels and employment effects: Implications for socio-economic development in Thailand. Biomass Bioenerg 46:409–418
25. Karekezi S, Kithyoma W (2006) Bioenergy and agriculture, promises and challenges. Bioenergy and the Poor. Vision, 2020 Focus 14, Brief 11 of 12, International Food Policy Research Institute, Washington, DC
26. De La Torre Ugarte DG (2006) Bioenergy and agriculture, promises and challenges. Developing Bioenergy: Economic and Social Issues. Vision, 2020 Focus 14, Brief 2 of 12, International Food Policy Research Institute, Washington, DC
27. Arshad M, Bano I, Younus M, Khan A, Rahman A (2017) Health concerns associated with biofuel production. Perspectives on water usage for biofuels production, pp. 97–105
28. Scovronick N, Wilkinson P (2014) Health impacts of liquid biofuel production and use: a review. Glob Environ Chang 24:155–164
29. Arbex MA, Martins LC, de Oliveira RC, Pereira LAA, Arbex FF, Cancado JED, Saldiva PHN, Braga ALF (2007) Air pollution from biomass burning and asthma hospital admissions in a sugar cane plantation area in Brazil. J Epidemiol Community Health 61(5):395–400
30. Wolf J, Bindraban P, Luijten J, Vleeshouwers L (2003) Exploratory study on the land area required for global food supply and the potential global production of bioenergy. Agric Syst 76(3):841–861
31. Baier S, Clements M, Griffiths C, Ihrig J (2009) Biofuels impact on crop and food prices: using an interactive spreadsheet. Board of Governors of the Federal Reserve System International Finance Discussion Papers Number 967

32. Ewing M, Msangi S (2009) Biofuels production in developing countries: assessing tradeoffs in welfare and food security. Environ Sci Policy 12(4):520–528
33. Thomson KJ (2009) The state of food and agriculture 2008: biofuels: prospects, risks and opportunities. Food and agriculture organisation of the united nations. J Agric Sci 147(4):503–503. (Cambridge University Press)
34. Arbex MA, Martins LC, de Oliveira RC, Pereira LAA, Arbex FF, Cancado JED, Saldiva PHN, Braga ALF (2010) Air pollution from biomass burning and asthma hospital admissions in a sugar cane plantation area in Brazil. J Epidemiol Community Health 61(5):395–400
35. Martin MA (2010) First generation biofuels compete. New Biotechnol 27(5):596–608
36. Graham-Rowe D (2011) Agriculture: beyond food versus fuel. Nature 474(7352):S6–S8
37. Pereira LG, Cavalett O, Bonomi A, Zhang Y, Warner E, Chum HL (2019) Comparison of biofuel life-cycle GHG emissions assessment tools: the case 2 studies of ethanol produced from sugarcane, corn, and wheat
38. FaistEmmenegger M, Pfister S, Koehler A, de Giovanetti L, Arena AP, Zah R (2011) Taking into account water use impacts in the LCA of biofuels: an Argentinean case study. Int J Life Cycle Assess 16(9):869–877
39. Rockström J, Williams J, Daily G, Noble A, Matthews N, Gordon L, Wetterstrand H, DeClerck F, Shah M (2016-05-13). Sustainable intensification of agriculture for human prosperity and global sustainability. Ambio 46(1):4–17

Current Developments in Bioremediation of Pesticides and Insecticides

S. Sivarathnakumar, R. Praveenkumar, J. Vinotharulraj, D. Gayathiri, and A. Amirthavarshini

Abstract Pests and insects are present in every nook and corner of the surrounding area which causes whooping economic losses. The substance or reagent that reduces the effective growth of these intruders like weeds, fungi, rodents and insects and other dreadful creatures are called pesticides. This chapter addresses the solutions to prohibit the toxic pollutants which spread in the pesticide form and it will become a priority for environmentalists and researchers working in this respective field.

Keywords Pesticides · Insecticides · Researchers

1 Introduction

Pesticides are chemical compounds that are used to kill pests including nematodes, insects and rodents. Fungicides, Herbicides and insecticides are also included in the domain of pesticides. Insecticides are used for preventing the spread of bacteria, it controls mice and rats, the role of herbicides is to destroy weeds, it controls unwanted vegetation, and the mould growth should be prevented by fungicides. Though pesticides and insecticides are controlling the widespread of pests and vectors, some residues remain in the topsoil and lead to toxicity by entering the food chain. Though the insecticide and pesticide residues are undergone various degradation processes in chemical, biochemical and physical form, some residues persist in the environment because of their water high stability and solubility. Mostly, the characteristics of soil under the environmental condition lead to persistence [1].

Whenever pesticides reach organisms, they could cause adverse effects. The level of toxicity depends upon the characteristics of the organism, environment and pesticide. Pesticide persistence leads to worse environmental impact by their physico-chemical properties [2].

S. Sivarathnakumar (✉)
Department of Chemical Engineering, Arunai Engineering College, Tiruvannamalai, India
e-mail: sivrathna@gmail.com

R. Praveenkumar · J. Vinotharulraj · D. Gayathiri · A. Amirthavarshini
Department of Biotechnology, Arunai Engineering College, Tiruvannamalai, India

N. A. Siddiqui et al. (eds.), *Advances in Waste Management*, Lecture Notes in Civil Engineering 301, https://doi.org/10.1007/978-981-19-7506-6_18

Insecticides play a major role in agricultural, medicinal and industrial fields but they cause an adverse effect in animals as well as in humans in toxic form, and it alters components in the ecosystem too. The concentric form of insecticides is easily entered into the food chain. This information could be used to eliminate unwanted effects by using organisms, and this approach has been called bioremediation [3].

Bioremediation is the best available option because of its eco-friendly nature and cost-effective benefits to the removal of pesticides from the environment. Bioremediation is a process where beneficial microbes like yeast, bacteria and fungi are used to clean the contaminants in water and soil. To recover and treat the environment, bioremediation is an attractive and effective technique in an eco-friendly manner [4]. By the application of biological methods, the pollutants and contaminants are eliminated or transformed. The major principle of bioremediation can either remove toxic substances or convert toxic to nontoxic pollutants with microbes usage [5]. Plant microbes are used in bioremediation techniques to remove pesticides from the soil more effectively. Basically, the remediation strategies should be addressed by synthetic biology for pesticides from the environment. The bioremediation efficiency should be increased by remediation strategies in microbial synthetic biology and it also provides methodologies for environmentalists and researchers. The bioremediation approaches should be furtherly developed and widely applied in the removal of toxic substances and other associated techniques.

This chapter will broadly discuss the recent development techniques in the bioremediation of pesticides and insecticides in summary. We hope that it will be helpful for more researchers to tackle the challenges of contamination of pesticides and insecticides.

2 Pesticides

Pesticides play a major role in crop production, but indiscriminate use of pesticides leads to health issues and risks in humans and causes severe pollution in the environment. Additionally, some residues of herbicides in the environment result in severe phytotoxicity. Therefore, pesticide elimination from soil crabs the attention of researchers [6].

2.1 Classification of Pesticides

Pesticides are compounds that occur naturally or synthetically produced, which are mostly poisonous, they can kill pests easily, and they may also be grouped to control the pest like insecticides, bactericides, herbicides, rodenticides, algicides, fungicides and nematicides. Pesticides like Atrazine, Endosulfan, benzene hexachloride (BHC), Dichlorodiphenyltrichloroethane (DDT), hexachlorocyclohexane (HCH), etc. are found in soil sediments because of their less bioavailability nature. The residues become unavailable to microorganisms because most of them get absorbed in the soil particles themselves. Based on the chemical class, the organic pesticides include triazoles, organophosphorus, organochlorine, neonicotinoids, triazines and pyrethroids. Among these, carbamates, organophosphates, organochlorines and synthetic-pyrethroids are gradually developed for their more effective nature [1].

Before the release of pesticides on the market, the active ingredients must be tested thoroughly and to evaluate the ecotoxicological effects if the risk is acceptable or not. Moreover, the toxicity level of pesticides on soil ecosystems is reported in various studies. These effects can fall into the following categories such as reduction of abundance microbial group, inhibition of microbial functions, diversity of soil microbial communities and changes in the composition. Organic chemical pesticides like organophosphates and organochlorine cause convulsions, irritability and tumours. Mostly, serious environmental issues are caused by organochlorine pesticides due to Biomagnification which threatens the whole ecosystem. Pesticides are mainly classified based on their toxicity, chemical nature and environmental persistence in the following groups [7].

2.1.1 Organochlorine Pesticides

Although organochlorine was used largely in the 1940s widely, due to its harmful effects and high persistence in the environment it was banned in developed countries in the 1970s. These pesticides have long residual effects, it has a lower level of toxicity, and they are hard to break down because of their chemical stability. This pesticide results in poisoning when it accumulates in mammals. Therefore, other pesticides should be used in place of organochlorine pesticide as it was banned. Some of the examples of organochlorine pesticides like endosulfan, DDT and lindane [8].

Endosulfan is one of the residues of pesticide which is more harmful to fishes and aquatic invertebrates, which blends with the soil and retains its persistence for a longer period. *Arthobacter* and *pseudomonas* species can degrade up to 56–92% of α-endosulfan and 73–95% β-Endosulfan in a period of 7 days [9].

2.1.2 Organophosphate Pesticides

Organophosphate pesticide act as nerve poison which is not only used as stomach poison but also as a fumigant and contact poison. In the environment, it causes lesser pollution, the resistance of pests is slow, and it is also easily biodegradable. Some of the organophosphate pesticides are Phosphomidone, methyl parathion and fenitrothion [10].

Table 1 Affecting factors and conditions required for pesticides

Factors	Conditions required
Nutrients	Carbon, Nitrogen, Oxygen, etc.
Microorganisms	Aerobic or Anaerobic
Soil moisture	25–28% of water holding capacity
Type of soil	Low clay or slit content
Environmental factors	Oxygen content Temperature, pH, Electron acceptor/donor
Biological methods of microorganisms	Catabolism and Anabolism

2.1.3 Carbamates

The principle of carbamates is similar to organophosphate pesticides where it affects the nerve signals of the pest and results in death. Mostly the carbamates are the same as natural organic substances when compare to their molecular structure, so they can be easily degradable with less pollution E.g.: Propoxur [11].

2.1.4 Synthetic-Pyrethroid Pesticides

By comparing the structure of natural pyrethrins, these pesticides are synthesized. Synthetic-pyrethroid pesticides are less poisonous to mammals and relatively high poisonous to insects. E.g.: Permethrin and Allethrin. The following factors are affecting the process of bioremediation of pesticides and the conditions required are tabulated below [source: Shanahan P Bioremediation, Waste Containment and Remediation Technology (2004), Springer] (Table 1).

2.2 Composition of Pesticides

The pesticides have many constituents, among them, the major compositions are charted below (Fig. 1).

2.3 History of Pesticides

Since 1940, synthetic organic pesticides such as hexachlorocyclohexane (BHC), dichloro-diphenyl-trichloroethane (DDT) and highly efficient chemical pesticides are used for controlling weeds, pathogens in plants and pests of agriculture where it done under large-scale production [12, 13]. Microbial degradation is the process that acts in a chemical pesticide metabolic pathway based on their advantages such

Fig. 1 Average composition of pesticides

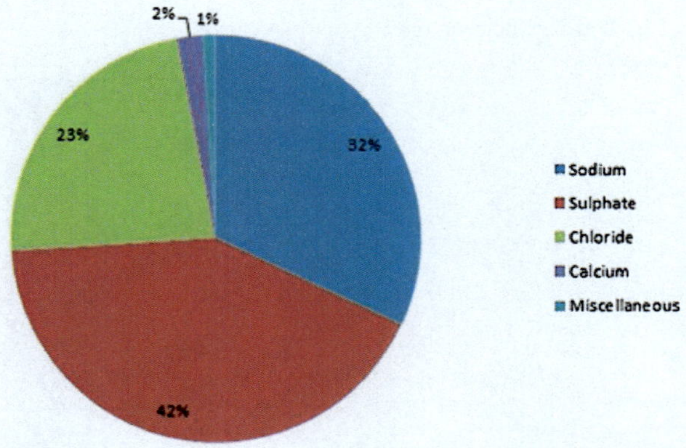

as cost-effective, environmental friendliness and high toxic removal efficiency in the environment (Xu et al., 2019). Microorganisms that particularly can degrade the numerous pesticides are identified and isolated successfully under different environmental conditions [14, 15].

Currently, many pesticides are eliminated from the market as they contain a high toxicity level [16]. The application of pesticides was caused many acute and chronic cases [17] and the level of resistance is increased in target species [18]. After that, the target species are replaced by the most harmful resistant species [19] and the various environmental conditions became contaminated [20, 21] According to the problem arisen, many researchers and higher authority peoples are demanded that the usage of pesticides should be appropriate to correct the legislation to protect the health of human and animal and risk of environment.

Recently, the concept of pesticides has been changed significantly to include high effectivity in the lower application, lesser toxicity to non-target species but higher selectivity to target organisms and lower persistence in the environment to avoid biomagnification and bioconcentration in the food chain to prohibit the resistance development. Hence new active substances are necessary to fight against pests effectively, but they have lesser effects on humans, animals and the environment.

3 Insecticides

More than targeted organism insecticides act on non-targeted organisms and these also affect humans and other aquatic creatures by mixing with water by leaching and whenever birds consume such water sources, they die due to the toxicity of the insecticide, such insecticide is DDT which affects the reproductive system of birds was banned in the United States (Fig. 2).

Insects may build resistance when they are continuously affected by insecticides and this takes place rapidly that the reproduction of insects takes place within three to four weeks so that the new generation with increased resistance is produced [22].

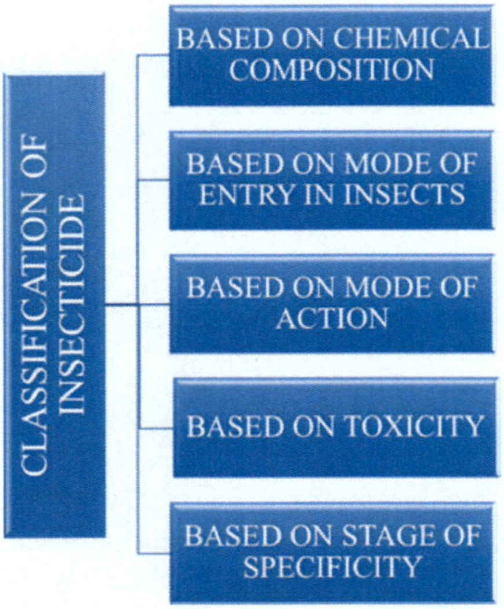

Fig. 2 Classification of insecticide

3.1 *Types*

Insecticides are of different types and they are

i. Systemic: The plant root absorbs these types of insecticide when they are introduced in soil and this moves to the external part of the plants such as fruits, leaves and branches. This insecticide forms a layer on the surface of the plant and when insects eat the plant, they affect the insects and kill them.
ii. Ingested: These pesticides are used to kill rats and roach.
iii. Contact: These are targeted insecticides that only kill specific insects. E.g. Households insect spray which directly targets the insects.

3.2 *Classification*

The insecticides are classified on a different basis and they are chemical composition: Organic and Inorganic pesticides.

i. Mode of Entry: These are poisons that affect insects and they are contact, fumigant and stomach poisons.
ii. Mode of action: These are chitin inhibitors and physical, nerve, respiratory, protoplasmic and general poisons.
iii. Toxicity: Extreme, moderate, High and less toxic insecticides.
iv. Stage of specificity: These affect the insects at all stages of growth such as ovicides, pupicides, larvicides that affect larvae and adulticides that affect the adult insects.

To reduce the pesticides and insecticides in the environment, Bioremediation is an effective way to transform toxic compounds into lesser or non-toxic substances by using living organisms. Recently, the studies have been gradually increased and focused on several mechanisms in the bioremediation process.

4 Methods of Bioremediation

The pesticides and insecticides waste and toxicity can be removed by use of bioremediation technique. Bioremediation also acts as a source of microbial growth by serving the contaminants as energy and food source. This technique is used for remediation of soil sludges and contamination of groundwater using solvents, wood preservatives, hydrocarbons of petrol, and other chemicals. The parameters such as Temperature, pH, Oxygen are controlled and required nutrients were added for microorganism growth. The rate of degradation of the contaminants by microorganisms is influenced by the presence of a concentration of contaminants and microbes, and various other required parameters such as temperature, pH, nutrient availability and supply, Moisture and co-metabolism. Based on the types of contamination specific microorganisms are added that enhance the bioremediation technique [23] Based on contaminants and process Bioremediation can be either done by In-situ or Ex-situ techniques.

4.1 In-Situ Bioremediation

The removal of toxic wastes and contaminants, where microbes use them as a feed and degrade those contaminants into non-toxic substances and dissolve the contaminants for biotransformation.

In-situ techniques are of two types

1. Intrinsic In-situ bioremediation
2. Engineered In-situ bioremediation

Intrinsic In-situ Bioremediation

This type of bioremediation technique uses naturally available microorganism in the surroundings and converts the environmental pollutants into non-toxic substances.

Engineered/Accelerated In-situ Bioremediation

This is a specific technique that accelerates the process by optimum growth of microbes under their optimum condition and Physio-chemical conditions of growth. Nutrients supply, Oxygen supply, and electron acceptors can increase the growth of microorganisms on the surface. This also includes Biostimulation and Bioaugmentation processes (Fig. 3).

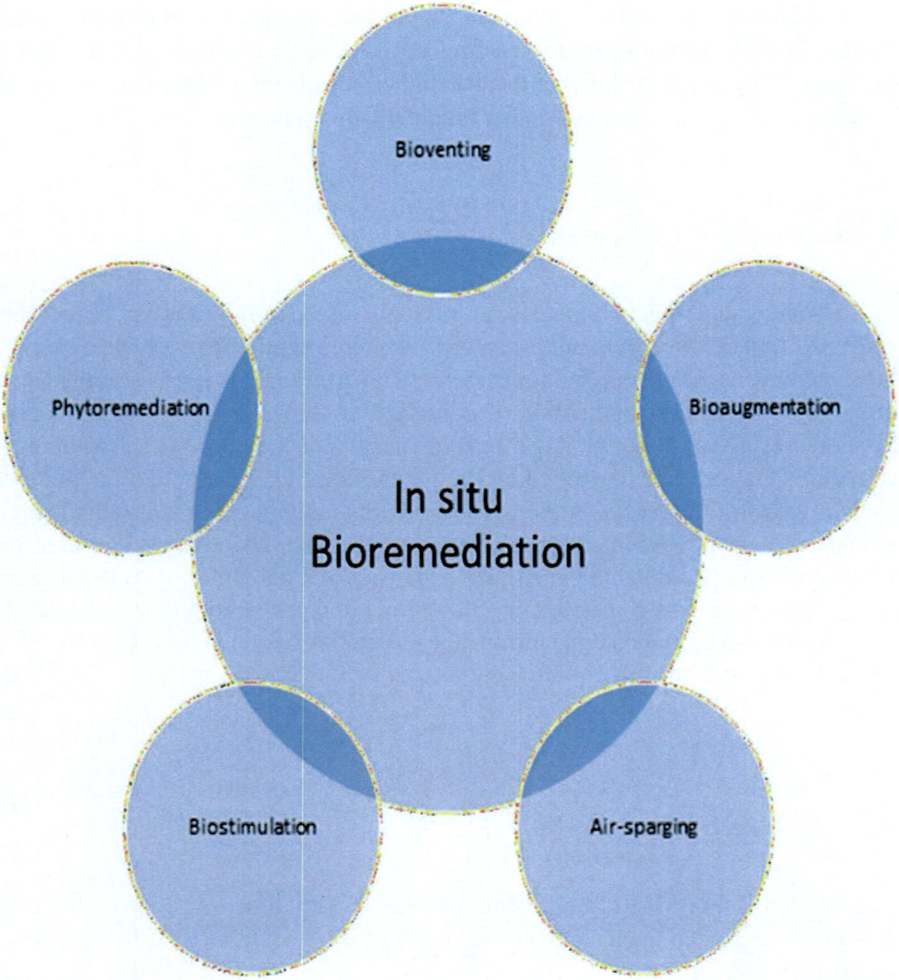

Fig. 3 Insitu bioremediation

4.1.1 Phytoremediation

The use of plant sources to remove the toxicity of chemicals and pollutants from the polluted areas is known as Phytoremediation. This is an eco-friendly approach that is also cost-effective for controlling environmental pollution.

4.1.2 Bioventing

Stimulation of *in-situ* bioremediation process of aerobically degradable pollutants presents in the soil by providing them more oxygen to an existing microorganism that enhances the process is called bioventing.

4.1.3 Bio Stimulation

Stimulation of growth of microorganism by mixing the contaminants with nutrients with vital components in gas or liquid form which results in quick contaminants removal using microbial population and this is the efficient process for removal of pollutants. While nutrients or food is supplied to that contaminated site, the parent microorganism population can survive and grow more of their intended work.

4.1.4 Bioaugumentation

This technique is used in special cases where the microorganisms are required to extract the contaminants like municipal wastewater and the major disadvantage of Bioaugmentation is impossible to control microorganism growth in process of removing the specific contaminants or pollutants [24].

Here, additional microorganisms are added to the contaminated site with pesticide so that this can be reinforced by the natural bioprocess.

4.1.5 Air Sparging

The mass removal of the contaminated saturated zone can be enhanced by the In-situ air sparging technique. This method is processed by injecting the air into the targeted contaminated zone. This technique is an exception for the removal of volatile and semi-volatile contaminants that may undergo mass transfer from the groundwater to bubbles.

4.2 Ex-situ Bioremediation

Ex-situ bioremediation is done by collecting the waste and contaminants substances for the polluted area that can assist the microbial growth and degradation of pollutants. The contaminants can either be in solid or liquid form. This technology is expensive as this technique includes solid handling, screening, excavation, homogenizing, mixing and disposal of final waste [25].

4.2.1 Land Farming

In Land farming, the soil which is contaminated is spread all over the land and tilled periodically which may improve the rate of aeration. The soil waste is initially excavated and then spread. The major advantage of this technique is cost- efficient as it reduced monitoring and maintenance.

4.2.2 Composting

The biological treatment process of solid-phase contaminants that are mixed with nontoxic organic amendments such as manure or other agricultural natural waste [26].

4.2.3 Bio Piles

Biopiles are the combination of two processes of land farming and composting at the same time and the advantage of this process is the physical loss of pollutants is controlled.

4.2.4 Bioreactors

From a contaminated site, water and soil are pumped up through an engineered containment system. Being the environment is more predictable and controllable, the degradation process in a bioreactor should be greater than in situ [27].

4.3 History of Bioremediation

The word "Bioremediation" means to destroy or immobilize the pollutant from the environment by using microorganisms.

Generally, Bioremediation is used for wastewater treatment. After many developments in this field, it is intentionally used for the reduction of hazardous waste. In 1960, George Robinson invented along the Santa Barbara coast that the oil spills were consumed by using microorganisms. He paved the way for modern bioremediation to consume pollutants by the use of microbes [28].

4.4 Advantages

While comparing with physical and chemical remediation methods, bioremediation is eco-friendly and has cost-saving features. The techniques of bioremediation make the preliminary stage less laborious and low cost in a short duration.

4.5 Limitations of Bioremediation

Despite several advantages, there are a few limitations also in the bioremediation method. If the environmental conditions are altered, then bioremediation may affect. Another very important limitation is the cost. It's very difficult for government and even for the industry to invest a huge amount of money in bioremediation. An additional problem is that environmental disruption may be caused by in-situ and ex-situ methods as these may cause damage to the environmental balance. The non-native bioremediation organisms should be introduced for long-term effects into the respective field are not at all gives positive result always [29].

5 Biopesticides in Ipm Practices

The usage of pesticides has many disadvantages including non-target species getting affected, development of resistances, water getting polluted, animal and human health getting risk. Even though pesticides have many advantages like disease control, many developed countries will demand to decrease the pesticides residue in the environment and in food. The pesticide usage should be compared with IPM (Integrated pest management) control measures to address these problems for current developments [30].

6 Current and Future Trends of Biopesticides

Biopesticides play a great role in resistance development and should be implemented in many programs. The agrochemical companies are important for new biopesticide development to control the activities and maintain the safety measures. The mode of action (MoA) is supported by providing funds that help in resistance management. The major goal of this system is to find the new biopesticide and their active substances with different MoA which involves the metabolic inhibition and lower the resistance risk. The market strategies and research and development of agrochemical companies might be changing because of pesticide resistance evolution and modern legislations should concentrate to protect the human, animal, and environmental health.

The European and Mediterranean Plant Protection Organization (EPPO) considers the "resistance risk analysis" (EPPO© 2015) it consists of 2 stages process "resistance risk assessment" and "resistance risk management". Where "resistance risk analysis" is used to demonstrate the pesticide formulations and active substances. The "resistance risk assessment" is used to examine the resistance development probability, meanwhile "resistance risk management" gives the strategy ideas to eliminate the resistance development. The resistance risk assessment needs to examine the agronomic risk (influenced by pattern, crop and geographic area) and the inherent risk (associated with the mode of action of the pest and product).

Many factors such as (active substance identification, level of resistance risk, cross-resistance, both resistance mechanism and MoA, resistant strains behaviour, unlimited usage consequences) are influence the process and the case studies should be examined. The performance of resistance risk management after resistance risk assessment stage should be done if the unacceptable risk is happening (high agronomic and inherent risk) or (low agronomic and inherent risk or even high inherent and low agronomic risk), the establishment of sensitivity baseline done through monitoring before the active substance introduction.

Finally, pesticide efficacy should be monitored to maintain a safe, sustainable and viable agricultural environment within the European Union. Therefore, avoid human, animal and environmental health risks by developing strategies in crop protection.

7 Future Perspective of Pesticide Bioremediation

Bioremediation is the technique for the pesticides and insecticides removal from contaminated water and soil, so it is emerging as the best option. Nowadays, various bioremediation approaches are available to address the problem of decontaminating the environmental compartments from toxic compounds. The various current bioremediation steps and their future perspective are as follows.

7.1 Gene Bioaugmentation

Recently, gene transfer has played a major role in microbial development and environmental adaptation [31]. In bioaugmentation, the most suggested technology is plasmid transformation through conjugation [32–35].

7.2 *Phytoaugmentation*

Phytoaugmentation means the remediation of genes should be added to microbial genes through plant engineering. The gene spreading in the environment should be controlled easily by gene incorporation into plants. The agricultural production should be done with insect resistance or herbicide engineering genes in most genetically engineered plants [36, 37].

7.3 *Immobilized Cells in Bioremediation*

Immobilized cells of recombinant *E.coli* used to detoxify the insecticides in the aqueous waste stream have been reported. Because of their ester bonds, these immobilized transgenic *E.coli* organisms could degrade many insecticides [38, 39].

7.4 *Rhizoremediation*

Rhizoremediation is a process of microbial gene characterization in pesticide rhizodegradation [40–42]. The various studies proved that the transfer of genes to indigenous bacteria from foreign bacteria easily degrades the pesticides and insecticides in the soil environment [43–45]. In transgenic plants, the ethylene-induced stress is overcome by its potential and rhizoremediation is improved by the production of biomass in roots and shoots [46, 47].

7.5 *Nanotechnology*

In recent years, Nanotechnology has become popular for its application in many sectors. The chlorinated organic compounds have been transformed effectively by iron-based nanomaterials. It is used in drinking water and beverage treatment which has traces of hazardous substances [48]. Gold and silver nanoparticles are usually capable of removing chlorpyrifos and malathion pesticides. These techniques should be preferred for the bioremediation of pesticides and insecticides.

8 Conclusion

Bioremediation in ecosystems is an eco-friendly and effective approach to ensure the efficient recycling of wastes and to preserve scarce resources. Bioremediation is an emerging field to reduce the pesticide and insecticide toxicity level to an undetectable amount. A genetic approach can elaborate on the perspectives of bioremediation. For novel methods, Gene editing tools should be used to degrading the toxic compounds.

The strategies of Biostimulation and Bioaugmentation are quite effective for pesticide decontamination. Moreover, immobilization technology and nanotechnology have shown remarkable potential in the field of bioremediation. In addition, the Rhizoremediation process along with the strategy of phytoaugmentation should employ GMO to have a bright future for the researchers.

References

1. Tarla DN, Erickson LE (2020) Phytoremediation and bioremediation of pesticide-contaminated soil. Appl Sci 2020(10):1217
2. Velazquez-Fernandez, Martinez-Rizo AB, Ramirez- Sandoval M (2012) Biodegradation and bioremediation of organic pesticides. Pesticides – Recent Trends Pestic Residue Assay 12:254–272
3. Divya M, Aanand S, Srinivasan A, Ahilan B (2015) Bioremediation – an eco- friendly tool for effluent treatment: a review. Int J Appl Res 1(12):530–537
4. Chawl N, Bhardwaj J, Singh L (2020) Bioremediation of organophosphate pesticides: current status and future prospective. Plant Arch 20:3405–3412
5. Nawaz K, Hussain K, Choudary N, Majeed A, Ilyas U, Ghani A, Lin F, Ali K, Afghan S, Raza G, Lashari MI (2011) Eco–friendly role of biodegradation against agricultural pesticides hazards. Afri J Microbiol Res 5(3):177–183
6. Huang X, He J, Yan X, Hong Q, Chen K, He Q et al (2017) Microbial catabolism of chemical herbicides: Microbial resources, metabolic pathways and catabolic genes. Pestic Biochem Physiol 143:272–297. https://doi.org/10.1016/j.pestbp.2016.11.010
7. Ahmad M, Ahmad I (2014) Bioremediation of pesticides: biotechnology, vol 11: biodegradation and bioremediation
8. Katayama A, Matsumura F (2009) Degradation of organochlorine pesticides, particularly endosulfan, by Trichoderma harzianum. Environ Toxicol Chem 12(6):1059–1065
9. Sutherland TD, Horne I, Harcourt RL, Russell RJ, Oakeshott JG (2002) Isolation and characterization of a Mycobacterium strain that metabolizes the insectide endosulfan. J Appl Microbiol 93:380–389
10. Singh BK, Kuhad RC, Singh A, Lal R, Tripathi KK (1999) Biochemical and molecular basis of pesticide degradation by microorganisms. Crit Rev Biotechnol 19:197–225
11. Parekh NR, Hartmann A, Charnay MP, Fournier JC (1995) Diversity of carbofuran–degrading soil bacteria and detection of plasmid–encoded sequences homologous to the mcd gene. FEMS Microbiol Ecol 17:149–160
12. Sharma R, Peshin R, Shankar U, Kaul V, Sharma S (2015) Impact evaluation indicators of an Integrated Pest Management program in vegetable crops in the subtropical region of Jammu and Kashmir, India. Crop Prot 67:191–199. https://doi.org/10.1016/j.cropro.2014.10.014
13. Sharma A, Shukla A, Attri K, Kumar M, Kumar P, Suttee A et al (2020) Global trends in pesticides: a looming threat and viable alternatives. Ecotoxicol Environ Saf 201:110812. https://doi.org/10.1016/j.ecoenv.2020.110812
14. Morya R, Salvachúa D, Thakur IS (2020) Burkholderia: an untapped but promising bacterial genus for the conversion of aromatic compounds. Trends Biotechnol 389:963–975. https://doi.org/10.1016/j.tibtech.2020.02.008
15. Arora PK (2020) Bacilli-mediated degradation of xenobiotic compounds and heavy metals. Front Bioeng Biotechnol 8:570307. https://doi.org/10.3389/fbioe.2020.570307
16. Howard LO (1935) La menace des insectes. Flammarion, Paris, France

17. Hunt EG, Bischoff AI (1960) Inimical effects on wildlife of periodic DDD applications to Clear Lake. Calif Fish Game 46:91–106
18. Mouches C, Pasteur N, Berge JB, Hyrien O, Raymond M, Desaintvincent BR, Desilvestri M, Georghiou GP (1986) Amplification of an esterase gene is responsible for insecticide resistance in a California Culex mosquito. Science 233(4765):778–780. https://doi.org/10.1126/science.3755546
19. Regnault-Roger C, Philogene BJR, Vincent C (2005) Biopesticides of plant origin. Intercept Ltd., Wimborne, UK
20. Leduc R, Unny TE, McBean EA (1987) Stochastic modeling of the insecticide fenitrothion in water and sediment compartments of a stagnant pond. Water Resour Res 23(7):1105–1112. https://doi.org/10.1029/WR023i007p01105
21. Ellgehausen H, Guth JA, Esser HO (1980) Factors determining the bioaccumulation potential of pesticides in the individual compartments of aquatic food-chains. Ecotox Environ Safe 4(2):134–157. https://doi.org/10.1016/0147-6513(80)90015-9
22. Oberemok VV, Zaitsev AS, Levchenko NN, Nyadar PM (2015) A brief review of most widely used modern insecticides and prospects for the creation of DNA insecticides. SSN 0013–8738. Entomol Rev 95(7): 824–831
23. Vaccari A, Strom F, Alleman E (2006) Environmental biology for engineers and scientists
24. Thierry L, Armelle B, Karine J (2008) Performance of bioaugmentation– assisted phytoextraction applied to metal contaminated soils: a review. Environ Pollut 153:497–522
25. Philip JC, Atlas RM (2005) Bioremediation of contaminated soils and aquifers. In: Atlas RM, Philp JC (eds) Bioremediation: applied microbial solutions for real-world environmental cleanup. American Society for Microbiology (ASM) Press, Washington, pp 139–236
26. Azubuike CC, Chikere CB, Okpokwasili GC (2016) World J Microbiol Biotechnol 32:180
27. Mohan SV, Sirisha K, Rao NC, Sarma PN, Reddy SJ (2004) Degradation of chlorpyrifos contaminated soil by bioslurry reactor operated in sequencing batch mode: bioprocess monitoring. J Hazard Mater 10:116(1–2):39–48
28. Uqab B, Mudasir S, Nazir R (2016) Review on bioremediation of pesticides. J Bioremediat Biodegrad 7:343. https://doi.org/10.4172/2155-6199.1000343
29. Mary Kensa V (2011) Bioremediation - an overview. J Indust Pollut Control 27(2):161–168
30. Xi X, Yan J, Quan G, Cui L (2014) Removal of the pesticide pymetrozine from aqueous solution by biochar produced from brewer's spent grain at different pyrolytic temperatures. BioResources 9:7696–7709
31. Ochman H, Lawrence JG, Groisman EA (2000) Lateral gene transfer and the nature of bacterial innovation. Nature 405:299
32. DiGiovanni GD, Neilson JW, Pepper IL, Sinclair NA (1996) Gene transfer of Alcaligenes eutrophus JMP134 plasmid pJP4 to indigenous soil recipients. Appl Environ Microbiol 62:2521
33. Top EM, Maila MP, Clerinx M, Goris J, De Vos P, Verstraete W (1999) Methane oxidation as a method to evaluate the removal of 2,4– dichlorophenoxyacetic acid (2,4–D) from soil by plasmid mediated bioaugmentation. FEMS Microb Ecol 28:203
34. Newby DT, Gentry TJ, Pepper IL (2000) Comparison of 2,4– dichlorophenoxyacetic acid degradation and plasmid transfer in soil resulting from bioaugmentation with two different pJP4 donors. Appl Environ Microbiol 66:3399
35. Dejonghe W, Goris J, El Fantroussi S, Hofte M, De Vos P, Verstraete W, Top EM (2000) Effect of dissemination of 2,4–dichlorophenoxyacetic acid (2,4–D) degradation plasmids on 2,4–D degradation and on bacterial community structure in two different soil horizons. Appl Environ Microbiol 66:3297
36. Perlak FJ, Oppenhuizen M, Gustafson K, Voth R, Sivasupramaniam S, Heering D, Carey B, Ihrig RA, Roberts JK (2001) Development and commercial use of Bollgard® cotton in the USA—Early promises versus today's reality. Plant J 27:489
37. Kota M, Daniell H, Varma S, Garczynski SF, Gould F, Moar WJ (1999) Overexpression of the Bacillus thuringiensis (Bt) Cry2Aa2 protein in chloroplasts confers resistance to plants against susceptible and Bt–resistant insects: Proc. Natl Acad Sci USA 96:1840

38. Kim SII, Kim SJ, Nam MH (2002) Proteome analysis of aniline– induced proteins in Acinetobacter lwoffi K24. Curr Microbiol 44:61–66
39. Huang J, Qiao Ch–L, Li, X. and Xing, J.M. (2001) Cloning and fusion expression of detoxifing gene in Escherichia coli. Acta Genet Sin 28:583–588
40. Shaw LJ, Burns RG (2004) Enhanced mineralisation of [U–14C] 2,4– dichlorophenoxyacetic acid in soil from the rhizosphere of Trifolium pretense. Appl Environ Microbiol 70:4766–4774
41. Shaw LJ, Burns RG (2005) Rhizodeposition and the enhanced mineralization of 2,4–dichlorophenoxyacetic acid in soil from the Trifolium pretense rhizosphere. Environ Microbiol 7:191–202
42. Piutti S, Hallet S, Rousseaux S, Philippot L, Soulas G, Martin– Laurent, F. (2002) Accelerated mineralisation of atrazine in maize rhizosphere soil. Biol. Fert. Soils 36:434–441
43. Zhang R, Cui Z, Zhang X, Jiang J, Gu JD, Li S (2006) Cloning of the organophosphorus pesticide hydrolase gene clusters of seven degradative bacteria isolated from a methyl parathion contaminated site and evidence of their horizontal gene transfer. Biodegradation 17:465–472
44. Devers M, Henry S, Hartmann A, Laurent FM (2005) Horizontal gene transfer of atrazine–degrading genes (atz) from Agrobacterium tumefaciens St96–4 pADP1:Tn5 to bacteria of maize–cultivated soil. Pest Manage. Sci. 61:870–880
45. Boltner D, Moreno-Morillas S, Ramos JL (2005) 16S rDNA phylogeny and distribution of lin genes in novel hexachlorocyclohexane-degrading Sphingomonas strains. Environ Microbiol 7:1329–1338
46. Arshad M, Hussain S, Saleem M (2007) Optimization of environmental parameters for biodegradation of alpha and beta endosulfan in soil slurry by Pseudomonas aeruginosa. J Appl Microbiol 104:364–370
47. Arshad M, Saleem M, Hussain S (2007) Perspectives of bacterial ACC deaminase in phytoremediation. Trends Biotechnol 8:356–362
48. Paknikar KM, Nagpal V, Pethkar AV, Rajwade JM (2005) Degradation of lindane from aqueous solutions using iron sulfide nanoparticles stabilized by biopolymers. Sci Technol Adv Mater 6:370–374